Thomas Kang

Hybridization of Structural Fibers for Synergistic & Ductile Behavior

AF 154070

Thomas Kang

Hybridization of Structural Fibers for Synergistic & Ductile Behavior

Tensile Test of Hybrid Carbon/Glass Fibers

LAP LAMBERT Academic Publishing

Impressum / Imprint

Bibliografische Information der Deutschen Nationalbibliothek: Die Deutsche Nationalbibliothek verzeichnet diese Publikation in der Deutschen Nationalbibliografie; detaillierte bibliografische Daten sind im Internet über http://dnb.d-nb.de abrufbar.

Alle in diesem Buch genannten Marken und Produktnamen unterliegen warenzeichen-, marken- oder patentrechtlichem Schutz bzw. sind Warenzeichen oder eingetragene Warenzeichen der jeweiligen Inhaber. Die Wiedergabe von Marken, Produktnamen, Gebrauchsnamen, Handelsnamen, Warenbezeichnungen u.s.w. in diesem Werk berechtigt auch ohne besondere Kennzeichnung nicht zu der Annahme, dass solche Namen im Sinne der Warenzeichen- und Markenschutzgesetzgebung als frei zu betrachten wären und daher von jedermann benutzt werden dürften.

Bibliographic information published by the Deutsche Nationalbibliothek: The Deutsche Nationalbibliothek lists this publication in the Deutsche Nationalbibliografie; detailed bibliographic data are available in the Internet at http://dnb.d-nb.de.

Any brand names and product names mentioned in this book are subject to trademark, brand or patent protection and are trademarks or registered trademarks of their respective holders. The use of brand names, product names, common names, trade names, product descriptions etc. even without a particular marking in this work is in no way to be construed to mean that such names may be regarded as unrestricted in respect of trademark and brand protection legislation and could thus be used by anyone.

Coverbild / Cover image: www.ingimage.com

Verlag / Publisher:
LAP LAMBERT Academic Publishing
ist ein Imprint der / is a trademark of
OmniScriptum GmbH & Co. KG
Bahnhofstraße 28, 66111 Saarbrücken, Deutschland / Germany
Email: info@lap-publishing.com

Herstellung: siehe letzte Seite /
Printed at: see last page
ISBN: 978-3-659-82957-4

Copyright © 2016 OmniScriptum GmbH & Co. KG
Alle Rechte vorbehalten. / All rights reserved. Saarbrücken 2016

TABLE OF CONTENT

1. INTRODUCTION

Since the late 1980's fiber-reinforced polymer (FRP) sheets or wraps have been used to replace corrosion-vulnerable steel plates in repair applications. FRP sheets offer the advantages of light weight, high strength, low cost, and constructability and durability (non-corrosiveness). Despite the expensive cost relative to glass fibers (GF), carbon fibers (CF) and carbon FRP sheets/plates have been primarily used for repair and retrofit. This is mainly due to the fact that CF has a high elastic modulus, high ultimate strength (see Figure 1), and superior durability. GF is also popular, particularly for column jacketing (confining) retrofit, as it only costs about 5 to 10% as much as CF. GF has much less ultimate stress and very low elastic modulus (only about a quarter of that of steel), but very large ductility (Figure 1). It is noted that aramid fibers (AF) have both very large ductility and relatively high elastic modulus (Figure 10); however, because AF is as costly as CF, little economic advantage may be gained from the use of AF.

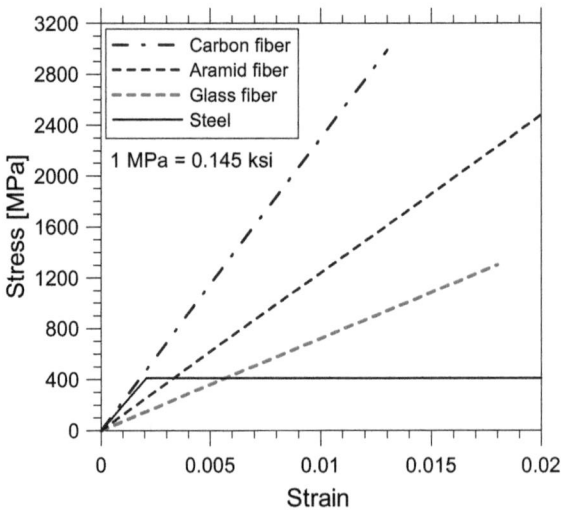

Figure 1 Comparison of Young's Moduli of Steel and Carbon, Aramid and Glass Fibers.

Brittleness is a major drawback of all these fibers (CF, AF and GF), since they have no yielding point and associated nonlinear behavior (Figure 1). To improve the ductility of the fibers, a number of composite material science investigations have been conducted on hybrid fibrous composites (e.g., [1], [2], [3], [4], [5], [6], [7], [8]). Applications of hybrid FRP composites, such as hybrid FRP bars and sheets combined with concrete, have been studied by several researchers (e.g., [9], [10], [11]). The primary purpose of these civil engineering applications was to achieve "pseudo-ductility" similar to the ductile response of nonlinear steel materials. Pseudo-ductility can be defined by the author of this book as when, after the first fiber failure (first drop in load), the load carrying capacity is recovered or improved as the remaining fibers elongate. Pseudo-ductility is also desirable because clear sound warning is produced during the first fiber failure, which indicates distress and the possible impending failure of structures.

The secondary purpose of hybrid FRP composites in structural engineering applications might be to actively utilize the so-called "hybrid effects." Marom et al. [7] defined the hybrid effects as the deviation of the behavior of a hybrid composite from the rule of mixtures, while Manders and Bader [2] simply defined it as the difference in behavior between a fiber in a hybrid composite and in a non-hybrid composite. Both positive and negative hybrid effects are possible; the effects are deemed positive (synergistic) when mechanical properties are above the prediction based on the rule of mixtures and vice versa for negative effects. It is extremely difficult to theoretically predict the hybrid effects and mechanical properties of hybrid fibrous composites, which are known to depend on the volumetric ratio of each fiber component, bonding property between the components, and elastic moduli of the fibers or their ratio ([8]). This is mainly due to the unavoidable uncertainty of the bonding property. Also, the size effect is involved. In structural engineering applications, hybrid FRP sheets or plates consisting of fiber rovings (strands) would be practical and feasible. A high-strength carbon fiber (CF) roving is typically made of about 12,000 filaments (12K) or multiples of 12,000 filaments (e.g., 24K or 48K), while an E-glass fiber (GF) roving is made of 1,200tex, 2200tex or multiple of 2,200tex, where 1tex is 1,000 m/g (or 459,920 yd/lb). Thus, some findings from previous research on a micro-composite or a composite made of a fraction of different fibers embedded in the composite matrix (i.e., in the fiber roving or strand) may not be applicable to the hybrid FRP sheets that are focused on structural repair or other structural engineering applications.

3

When the hybrid carbon-glass FRP sheet is subjected to tension, the CF with high elastic modulus and low ultimate strain ruptures first. The GF, with lower elastic modulus and higher ultimate strain, then takes over and resists the load. As noted, if the stress at GF rupture is equal to or higher than that at CF rupture, which depends on a volume ratio of (GF/CF) (e.g., [2]), the pseudo-ductility can be obtained. Hybrid effects are also expected to be gained, such that it is possible to enhance (first) failure stress (or strain) beyond that predicted from the rule of mixtures, given Equation (1) below:

$$E_{HF} = E_{CF}\left(\frac{V_{CF}}{V_{HF}}\right) + E_{GF}\left(\frac{V_{GF}}{V_{HF}}\right) \tag{1}$$

where E_{HF} is the weighted mean elastic modulus of a carbon-glass hybrid composite; E_{CF} and E_{GF} are the elastic moduli of CF and GF, respectively; V_{CF} and V_{GF} are the CF and GF volumes, respectively; and V_{HF} is the combined CF and GF volume or the volume of the hybrid composites.

Manders and Bader [2] reported that the increase in strain at CF rupture in sandwich laminated hybrids would be about 50% of that of single CF, and Aveston and Sillwood [6] also experimentally confirmed that the strain at CF rupture of hybrid carbon-glass-epoxy composites could be increased up to about 0.01. Furthermore, Miwa and Horiba [4] suggested the empirical rule of "hybrid" mixtures as:

$$f_{u\_C\_HF} = f_{u\_CF}\left(\frac{V_{CF}}{V_{HF}}\right) + f_{u\_GF}\left(\frac{V_{GF}}{V_{HF}}\right) \tag{2}$$

where $f_{u\_C\_HF}$ is the mean stress of a carbon-glass hybrid composite at CF rupture and f_{u_CF} and f_{u_GF} are the ultimate stresses of CF and GF ruptures, respectively.

However, researchers (e.g., [2], [4]) did not reach any definite conclusion on the ultimate stress of hybrid carbon-glass composites at GF rupture. Pan and Postle [8] reported that due to the cross-coupling effects between the different fibers, a positive hybrid effect would be expected at the first fiber rupture, whereas a negative effect would be expected at the second fiber rupture; however, this appears to be the case only for the first fiber embedded in the matrix or a postulate without examination of an optimal ratio of two different fibers. It is not appropriate to apply the shear lag model ([12]) to the case of interest, since the hybrid sheet may have a substantially different degree of interfacial shear stress as in the case of a short-fiber embedded in the matrix. The increased or decreased strain (or stress) at GF rupture of the hybrid composites, particularly hybrid FRP sheets that are common in civil engineering applications, have not been well studied. A continuous FRP sheet consisting of fiber rovings may have a moderate level of frictional coupling between GF and CF rovings and behave very differently than the micro-composites with a high level of frictional coupling.

The current study consists of in-depth re-assessment of two hybrid material test programs conducted by collaborators ([13], [14], [15]), who generously provided test raw data, identification of hybrid effects in the carbon-glass FRP sheets, and development of design models for stress-strain relationships with and without consideration of the hybrid effects. It is specifically noted that the author of this report conducted a detailed analysis of all raw data that were provided by Dr. Chin Yong Lee [14] and re-examined the results from the raw data to gain a newer and better understanding of the behavior of the hybrid FRP sheets. This book includes the documentation of this research process and findings.

2. MATERIAL TEST PROGRAMS

Two independent material test programs on uniaxial hybrid fiber-reinforced polymer (FRP) sheets and each corresponding fiber used for the fabrication of the hybrid sheets are presented in this chapter. Note that although ACI 440.2R-08 [16], Section 4.3.1 recommends using mean minus three times standard deviations for the ultimate stress and strain from at least 20 samples, the number for the tested sample was less than 20 for all three test programs.

2.1 FIRST TEST PROGRAM

Choi et al. [13] tested rovings of high-strength carbon fibers (CF) and E-glass fibers (GF) in tension, hybrid carbon-glass FRP sheets in tension, and epoxy adhesives in bending (J type). Both bare and impregnated rovings were tested, and digital data of forces and displacements at ultimate indicated by a Universal Test Machine (U.T.M; Lloyd Instruments, LR5K) with a 500 N (112 lbs) capacity were manually recorded. The cross-sectional areas of CF and GF rovings were 0.886 and 0.444 mm^2 (0.0014 and 0.0007 in^2), respectively. This is based on each material's Specific Gravity (ρ_{CF} = 1.8; ρ_{GF} = 2.54), and the measured weight and length. The impregnated roving tests were carried out in accordance with ASTM D3039-08 [17]. The test specimens had a total length of 400 mm (15.75 in.) and an effective length between yarn grips of 260 mm (10.25 in.). The tensile loading speed was 10 mm/min (0.4 in./min).

For the bar roving tests, a pair of thin, flat plastic films were laminated in the grips. For the impregnated roving tests, two different types of grips (Types A and B) were developed. The Type A grip tabs were applied using an epoxy adhesive. According to ASTM D3039-08 [17], no industry consensus on the grip at the end of the fiber coupon is available. Thus, although grip failures were not observed from any of the methods used in the test program ([13]), an alternative method of gripping was developed, specifically, 90° sandwich laminates using the same fiber rovings (Type B grip). Table 1 summarizes the average of the results from rovings categorized as non-impregnated, impregnated-Type A grip and impregnated-Type B grip. As shown, there are no substantial discrepancies between the methods and within a method (standard deviations were relatively small), indicating that all of the test methods yield essentially consistent results.

Table 1 Measured Results for the Uniaxial Tensile Tests of CF And GF Rovings Conducted by Choi et al. (2009) [13] and Provided Properties of CF and GF Filaments

	Carbon filament		Glass filament
σ_{u_CF}	4,900	σ_{u_GF}	2,900
ε_{u_CF}	0.0213	ε_{u_GF}	0.0401
E_{CF}	230	E_{GF}	72.4

	CF-N-1	CF-N-2	CF-N-3	CF-N-4	CF-N-5	CF-N-6	CF-N-7			Average
σ_{u_CF}	1,248	1,529	1,273	1,241	1,069	1,406	1,208			1,283
ε_{u_CF}	0.011	0.0123	0.0117	0.0114	0.01	0.0125	0.0105			0.0113
E_{CF}	114	125	109	109	107	113	115			113

	CF-A-1	CF-A-2	CF-A-3	CF-B-1	CF-B-2	CF-B-3	CF-B-4	CF-B-5	CF-B-6	Average
σ_{u_CF}	1,734	1,680	1,382	1,628	1,544	1,354	1,515	1,533	1,624	1,553

	GF-N-1	GF-N-2	GF-N-3	GF-N-4	GF-N-5	GF-N-6	GF-N-7	GF-N-8	GF-N-9	GF-N-10	Average
σ_{u_GF}	708	883	757	776	901	769	801	794	785	742	792
ε_{u_GF}	0.0167	0.0185	0.0177	0.0167	0.0209	0.0191	0.0174	0.0166	0.0154	0.0169	0.0176
E_{GF}	42.3	47.8	42.9	46.4	43.2	40.3	46.0	47.8	51.1	43.9	45.1

	GF-A-1	GF-A-2	GF-A-3	GF-A-4	GF-A-5	GF-B-1	GF-B-2	GF-B-3	GF-B-4	GF-B-5	Average
σ_{u_GF}	580	584	869	571	661	1052	951	890	799	681	764

Filament properties are provided by the manufacturer.
CF = Carbon fiber roving; GF = Glass fiber roving.
N = Non-impregnated rovings; A = Impregnated-Type A grip; B = Impregnated-Type B grip.
σ_{u_CF} Measured ultimate stress of CF roving [MPa].
ε_{u_CF} Measured ultimate strain of CF roving.
E_{CF} Measured elastic modulus of CF roving [Gpa].
σ_{u_GF} Measured ultimate stress of GF roving [Mpa].
ε_{u_GF} Measured ultimate strain of GF roving.
E_{GF} Measured elastic modulus of GF roving [Gpa].
Conversion: 1 Mpa = 0.145 ksi; 1 Gpa = 145 ksi.

Carbon-glass hybrid FRP sheets were fabricated with a (GF/CF) volumetric ratio of (8.8/1) (see Figure 2), and tested in tension in accordance with CSA S806-02 [18]. Table 2 indicates the averaged values for three samples of the hybrid FRP sheets. Each sample has a cross-sectional area of about 17.5 mm^2 (0.027 in^2), with a 48K-CF roving of 1.8 mm^2 (0.0028 in^2) and eighteen 2,200tex-GF rovings of 15.6 mm^2 (0.024 in^2). Both strain gauges and LVDTs embedded in the U.T.M. (Instron) with a capacity of 1,200 kN (270 kips) were used to digitally monitor stains or displacement. In this study, the LVDT data were more reliable than the strain gauge data.

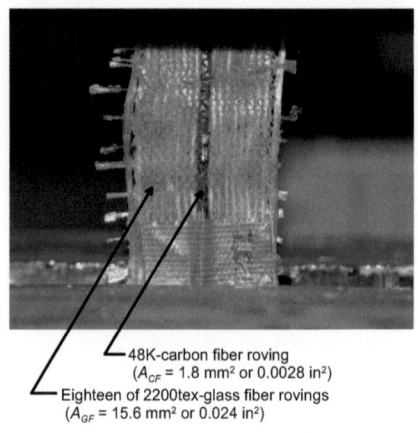

48K-carbon fiber roving
(A_{CF} = 1.8 mm^2 or 0.0028 in^2)
Eighteen of 2200tex-glass fiber rovings
(A_{GF} = 15.6 mm^2 or 0.024 in^2)

Figure 2 Tensile Testing of Hybrid Carbon-Glass FRP Sheet (Choi et al., 2009 [13])

In order to identify mechanical properties of epoxy adhesives, flexural tests were conducted instead of direct tensile tests. This is because, first, the flexural tests are much more convenient, and second, a tensile strain is generally smaller than the actual strain of the adhesive bonded to concrete (ASTM D638 [19]). In this study, epoxy solid blocks (J type) with dimensions of 25 x 25 x 240 mm (1 x 1 x 9.5 in.) were tested in flexure under four-point loading. The resulting average values of ultimate stress and strain and modulus of elasticity at rupture for three specimens are 42.8 Mpa (6.2 ksi), 0.0197 and 2.19 GPa (317.6 ksi), respectively, which are similar to the typical values reported by ACI 503R-93 [20]. The ultimate strain of the product was relatively low. It would be useful to have an ultimate strain of about 0.04 to ensure that fiber ruptures prior to epoxy failure.

8

Table 2 Measured and Predicted Results for the Uniaxial Tensile Tests of
Hybrid FRP Sheets with (GF/CF = 8.8/1) Conducted by Choi et al.
(2009) [13]

Specimen	Meas. E_{HF} [Gpa]	Meas. $\varepsilon_{u\_C\_HF}$	Meas. $\sigma_{u\_C\_HF}$ [Mpa]	$E_{HF}^{\dagger}$ [Gpa]	$\varepsilon_{u\_C\_HF}^{\dagger}$	$\sigma_{u\_C\_HF}^{\dagger}$ [Mpa]	$\sigma_{u\_C\_HF}^{\ddagger}$ [Mpa]	Meas. E_{GF}^{*} [Gpa]	Meas. $\varepsilon_{u\_G\_HF}$	Meas. $\sigma_{u\_G\_HF}$ [Mpa]	$\sigma_{u\_G\_HF}^{\dagger}$ [Mpa]
(1)	(2)	(3)	(4)	(5)	(6)	(7)	(8)	(9)	(10)	(11)	(12)
Hybrid-1	59.2	0.0153	864.5	52	0.011	583.9	841.8	59	0.0218	1,040	711
Hybrid-2	67.1	0.0144	864.5	52	0.011	583.9	841.8	51.2	0.0181	928	711
Hybrid-3	68.5	0.0152	876.3	52	0.011	583.9	841.8	54	0.0228	1,277	711
Average	64.9	0.0150	868.4	52	0.011	583.9	841.8	54.7	0.0209	1,082	711

Meas. = Measured.
† = based on the rule of mixtures.
‡ = based on the rule of hybrid mixtures (Miwa and Horiba, 1994).
E_{HF} Elastic modulus of hybrid FRP sheet.
$\sigma_{u\_C\_HF}$ Stress at CF rupture of hybrid FRP sheet.
$\varepsilon_{u\_C\_HF}$ Strain at CF rupture of hybrid FRP sheet.
E_{GF}^{*} Average stress increase of hybrid FRP sheet divided by strain increase after CF rupture.
$\sigma_{u\_G\_HF}$ Stress at GF rupture of hybrid FRP sheet.
$\varepsilon_{u\_G\_HF}$ Strain at GF rupture of hybrid FRP sheet.
Conversion: 1 Mpa = 0.145 ksi; 1 Gpa = 145 ksi.

2.2 SECOND TEST PROGRAM

Dr. Chin Yong Lee's team [14] conducted tensile tests of conventional carbon and glass
FRP sheets and hybrid carbon-glass FRP sheets with a variety of (GF/CF) ratios ranging
from (1/1) to (10/1). The team generously provided raw data to the author of this book,
who re-analyzed and re-assessed the data completely as shown in the Appendix. Each
sheet had a total length of about 250 mm (10 in.) (Figure 3). High-strength carbon fiber
(CF) and E-glass fiber (GF) rovings were used along with two different types of epoxy
resins (J type and K type) to fabricate a total of 99 impregnated hybrid sheet coupons. The
J type epoxy is commonly used in practice, and the K type epoxy has a higher ductility.
Tensile tests of the epoxy resins turned out to be infeasible due to the grip problem; thus,
four-point loading tests were conducted on 25 x 25 x 240 mm (1 x 1 x 9.5 in.) molded

epoxy blocks. The ultimate strains for the J and K types were measured to be 0.02 and 0.029, respectively.

Using Specific Gravity (ρ_{CF} = 1.8 and ρ_{GF} = 2.54) and measured weights and lengths, the cross-sectional areas of CF and GF rovings were calculated to be 0.45 and 0.455 mm^2 (0.0007 and 0.000705 in^2), respectively. Controlling the number of each fiber roving (12K-CF roving and 1,200tex-GF roving), hybrid FRP sheets with 12 different (GF/CF) ratios were made (Table 3), including carbon FRP and glass FRP sheets. However, all the data from the glass FRP sheets were misplaced. Each sheet was impregnated with epoxy resin in a mold, where Overhead Projector (OHP) films were used to make the sample detachable from the mold. The width of the sample ranged from 11.9 to 16.5 mm (0.47 to 0.65 in.), and the strip thickness was 1.5 mm (0.06 in.). An epoxy-to-fiber ratio of 1.5 was used.

Tension forces were digitally recorded from the Universal Test Machine (U.T.M), tensile strains from the strain gauges mounted on the impregnated FRP sheet and total elongations from the LVDTs. The tensile loading speed was 1 mm/min (0.04 in./min).

Table 3 Measured and Predicted Results for the Uniaxial Tensile Tests Conducted by Dr. Chin Yong Lee's team [14]

Specimen	Meas. E_{HF} [GPa]	Meas. $\varepsilon_{u\_C\_HF}$	Meas. $\sigma_{u\_C\_HF}$ [MPa]	$E_{HF}^\dagger$ [GPa]	$\varepsilon_{u\_C\_HF}^\dagger$	$\sigma_{u\_C\_HF}^\dagger$ [MPa]	$\sigma_{u\_C\_HF}^\ddagger$ [MPa]	Meas. E_{GF}^* [GPa]	Meas. $\varepsilon_{u\_G\_HF}$	Meas. $\sigma_{u\_G\_HF}$ [MPa]	$\sigma_{u\_G\_HF}^\dagger$ [MPa]
(1)	(2)	(3)	(4)	(5)	(6)	(7)	(8)	(9)	(10)	(11)	(12)
CFRP-a	273.8	0.01	2,425	202	0.0129	2,656	3,474				
CFRP-b	196	0.0125	2,450	202	0.0129	2,656	3,474				
CFRP-c	189.7	0.0146	2,597	202	0.0129	2,656	3,474				
CFRP-d	238.4	0.0105	2,496	202	0.0129	2,656	3,474				
CFRP-e	194	0.0145	2,660	202	0.0129	2,656	3,474				
CFRP-D-a	121.3	0.0108	2,404	202	0.0129	2,656	3,474				
CFRP-D-b	233.7	0.0153	3,555	202	0.0129	2,656	3,474				
CFRP-D-c	172.9	0.0154	2,662	202	0.0129	2,656	3,474				
Average	202.4	0.013	2,656	202	0.0129	2,656	3,474				
HFRP-1-a	130.9	0.0157	2,057	134.7	0.0129	1,738	1,920				592
HFRP-1-b	164.9	0.0125	1,815	134.7	0.0129	1,738	1,920				592
HFRP-1-c	145.3	0.0152	2,202	134.7	0.0129	1,738	1,920	244.3	0.0173	4,473	592

Specimen	Meas. E_{HF} [GPa]	Meas. $\varepsilon_{u\_C\_HF}$	Meas. $\sigma_{u\_C\_HF}$ [MPa]	$E_{HF}^\dagger$ [GPa]	$\varepsilon_{u\_C\_HF}^\dagger$	$\sigma_{u\_C\_HF}^\dagger$ [MPa]	$\sigma_{u\_C\_HF}^\ddagger$ [MPa]	Meas. E_{GF}^* [GPa]	Meas. $\varepsilon_{u\_G\_HF}$	Meas. $\sigma_{u\_G\_HF}$ [MPa]	$\sigma_{u\_G\_HF}^\dagger$ [MPa]
HFRP-1-d	148	0.0179	2,644	134.7	0.0129	1,738	1,920				592
HFRP-1-e	156.5	0.0167	2,384	134.7	0.0129	1,738	1,920				592
HFRP-1-f	136.9	0.0142	1,945	134.7	0.0129	1,738	1,920				592
HFRP-1-g	130.9	0.0157	2,057	134.7	0.0129	1,738	1,920				592
HFRP-D1-a	150.5	0.0135	2,027	134.7	0.0129	1,738	1,920	208.9	0.0151	4,234	592
HFRP-D1-b	144.6	0.0133	1,873	134.7	0.0129	1,738	1,920	238.7	0.016	4,030	592
HFRP-D1-c	150.7	0.0134	2,077	134.7	0.0129	1,738	1,920				592
Average	146	0.0148	2,108	134.7	0.0129	1,738	1,920	230.6	0.0161	4,246	592
HFRP-2-a	259.8	0.0053	1,477	112.2	0.0129	1,447	1,675				790
HFRP-2-b	99.4	0.0156	1,578	112.2	0.0129	1,447	1,675	143.8	0.0186	2,014	790
HFRP-2-c	100.8	0.0135	1,357	112.2	0.0129	1,447	1,675	99.1	0.017	2,097	790
HFRP-2-d	100.9	0.0132	1,328	112.2	0.0129	1,447	1,675				790
HFRP-2-e	92.8	0.0154	1,427	112.2	0.0129	1,447	1,675	126.8	0.0161	1,589	790
HFRP-D2-a	126.7	0.0163	2,066	112.2	0.0129	1,447	1,675				790
HFRP-D2-b	121.3	0.0187	2,267	112.2	0.0129	1,447	1,675				790
HFRP-D2-c	106.5	0.018	1,756	112.2	0.0129	1,447	1,675				790
Average	126	0.0145	1,657	112.2	0.0129	1,447	1,675	123.2	0.0172	1,900	790
HFRP-3-a	106.3	0.0191	2,039	101	0.0129	1,303	1,552				888
HFRP-3-b	89.1	0.0095	847	101	0.0129	1,303	1,552				888
HFRP-3-c	114.7	0.0135	1,553	101	0.0129	1,303	1,552				888
HFRP-3-d	117.5	0.013	1,527	101	0.0129	1,303	1,552	65.5			888
HFRP-3-e	111.8	0.0168	1,883	101	0.0129	1,303	1,552	111.8	0.0195	2,447	888
HFRP-3-f	91.6	0.0166	1,488	101	0.0129	1,303	1,552				888
HFRP-3-g	97.4	0.0124	1,203	101	0.0129	1,303	1,552				888
HFRP-D3-a	109.4	0.0171	1,875	101	0.0129	1,303	1,552				888
HFRP-D3-b	92.9	0.0164	1,590	101	0.0129	1,303	1,552				888
HFRP-D3-c	104.1	0.0184	1,914	101	0.0129	1,303	1,552				888
Average	103.5	0.0153	1,592	101	0.0129	1,303	1,552	88.7	0.0195	2,447	888
HFRP-4-a	107.5	0.0178	1,968	94.2	0.0129	1,215	1,479				948
HFRP-4-b	93.6	0.0185	1,729	94.2	0.0129	1,215	1,479				948
HFRP-4-c	88.3	0.021	1,854	94.2	0.0129	1,215	1,479				948
HFRP-4-d	101.6	0.0191	1,939	94.2	0.0129	1,215	1,479				948
HFRP-4-e	108.7	0.016	1,725	94.2	0.0129	1,215	1,479				948
HFRP-4-f	87.3	0.0193	1,684	94.2	0.0129	1,215	1,479				948
HFRP-4-g	96.9	0.0183	1,694	94.2	0.0129	1,215	1,479				948
HFRP-D4-a	98.7	0.0178	1,757	94.2	0.0129	1,215	1,479				948
HFRP-D4-b	104.5	0.0185	1,936	94.2	0.0129	1,215	1,479				948
HFRP-D4-c	103.6	0.0172	1,889	94.2	0.0129	1,215	1,479				948
Average	99.1	0.0184	1,818	94.2	0.0129	1,215	1,479				948
HFRP-5-a	77.9	0.0161	1,253	89.4	0.0129	1,153	1,426				990
HFRP-5-b	79.1	0.0197	1,556	89.4	0.0129	1,153	1,426				990
HFRP-5-c	31.9	0.0174	554	89.4	0.0129	1,153	1,426				990
HFRP-5-d	78.2	0.0178	1,362	89.4	0.0129	1,153	1,426				990
HFRP-5-e	80.3	0.0201	1,618	89.4	0.0129	1,153	1,426				990
HFRP-D5-a	69.8	0.0192	1,834	89.4	0.0129	1,153	1,426				990

11

Specimen	Meas. E_{HF} [GPa]	Meas. $\varepsilon_{u\_C\_HF}$	Meas. $\sigma_{u\_C\_HF}$ [MPa]	$E_{HF}^{\dagger}$ [GPa]	$\varepsilon_{u\_C\_HF}^{\dagger}$	$\sigma_{u\_C\_HF}^{\dagger}$ [MPa]	$\sigma_{u\_C\_HF}^{\ddagger}$ [MPa]	Meas. E_{GF}^{*} [GPa]	Meas. $\varepsilon_{u\_G\_HF}$	Meas. $\sigma_{u\_G\_HF}$ [MPa]	$\sigma_{u\_G\_HF}^{\dagger}$ [MPa]
HFRP-D5-b	97.2	0.018	1,753	89.4	0.0129	1,153	1,426				990
HFRP-D5-c	99	0.0185	1,834	89.4	0.0129	1,153	1,426				990
Average	76.7	0.0184	1,471	89.4	0.0129	1,153	1,426				990
HFRP-6-a	74.4	0.0185	1,375	86.3	0.0129	1,113	1,392	71.2			1,018
HFRP-6-b	80.9	0.0175	1,417	86.3	0.0129	1,113	1,392				1,018
HFRP-6-c	80.9	0.0177	1,430	86.3	0.0129	1,113	1,392				1,018
HFRP-6-d	80.6	0.018	1,450	86.3	0.0129	1,113	1,392	84.1	0.0193	1,726	1,018
HFRP-6-e	78.5	0.0199	1,559	86.3	0.0129	1,113	1,392				1,018
HFRP-6-f	81.2	0.0193	1,565	86.3	0.0129	1,113	1,392				1,018
HFRP-D6-a	82.3	0.0178	1,465	86.3	0.0129	1,113	1,392	53.6	0.0233	1,511	1,018
HFRP-D6-b	92.6	0.0207	1,783	86.3	0.0129	1,113	1,392				1,018
HFRP-D6-c	93.8	0.0188	1,767	86.3	0.0129	1,113	1,392				1,018
Average	82.8	0.0187	1,535	86.3	0.0129	1,113	1,392	69.6	0.0213	1,619	1,018
HFRP-7-a	68.2	0.0193	1,313	83.9	0.0129	1,082	1,366				1,038
HFRP-7-b	73.6	0.0173	1,269	83.9	0.0129	1,082	1,366				1,038
HFRP-7-c	61.9	0.0166	1,027	83.9	0.0129	1,082	1,366				1,038
HFRP-7-d	78.6	0.0158	1,248	83.9	0.0129	1,082	1,366				1,038
HFRP-7-e	66.8	0.0178	1,192	83.9	0.0129	1,082	1,366				1,038
HFRP-7-f	68	0.021	1,604	83.9	0.0129	1,082	1,366				1,038
HFRP-7-g	81.6	0.0176	1,464	83.9	0.0129	1,082	1,366	62.3	0.02	1,486	1,038
HFRP-D7-a	87.3	0.0193	1,682	83.9	0.0129	1,082	1,366	118.3	0.0229	1,918	1,038
HFRP-D7-b	91.6	0.0203	1,863	83.9	0.0129	1,082	1,366				1,038
HFRP-D7-c	79.7	0.0205	1,645	83.9	0.0129	1,082	1,366				1,038
Average	75.7	0.0186	1,431	83.9	0.0129	1,082	1,366	90.3	0.0215	1,702	1,038
HFRP-8-a	71.4	0.018	1,286	82.1	0.0129	1,059	1,346				1,054
HFRP-8-b	73.1	0.0211	1,543	82.1	0.0129	1,059	1,346				1,054
HFRP-8-c	71.3	0.0173	1,236	82.1	0.0129	1,059	1,346				1,054
HFRP-8-d	74.9	0.02	1,454	82.1	0.0129	1,059	1,346	40.1			1,054
HFRP-8-e	73.7	0.0184	1,356	82.1	0.0129	1,059	1,346	58.2	0.0233	1,359	1,054
HFRP-8-f	88	0.0206	1,811	82.1	0.0129	1,059	1,346	56	0.0251	1,772	1,054
HFRP-8-g	79.8	0.0221	1,766	82.1	0.0129	1,059	1,346	49			1,054
HFRP-D8-a	85.4	0.0179	1,529	82.1	0.0129	1,059	1,346				1,054
HFRP-D8-b	82.2	0.0184	1,514	82.1	0.0129	1,059	1,346				1,054
HFRP-D8-c	89.8	0.0141	1,263	82.1	0.0129	1,059	1,346				1,054
Average	79	0.0188	1,476	82.1	0.0129	1,059	1,346	50.8	0.0242	1,566	1,054
HFRP-9-a	84.2	0.0211	1,777	80.6	0.0129	1,040	1,330	92.8	0.0245	1,766	1,067
HFRP-9-b	87.2	0.0211	1,837	80.6	0.0129	1,040	1,330	107.2	0.0235	1,833	1,067
HFRP-9-c	89.1	0.0203	1,807	80.6	0.0129	1,040	1,330	79.5	0.0251	1,841	1,067
HFRP-9-d	101.8	0.0193	1,697	80.6	0.0129	1,040	1,330				1,067
HFRP-9-e	99.4	0.0171	1,717	80.6	0.0129	1,040	1,330	53.8	0.0233	1,865	1,067
HFRP-9-f	94.5	0.0192	1,986	80.6	0.0129	1,040	1,330	67.2			1,067
HFRP-9-g	78.5	0.0201	1,580	80.6	0.0129	1,040	1,330				1,067
HFRP-D9-a	84.4	0.0187	1,645	80.6	0.0129	1,040	1,330	36.2	0.0192	1,715	1,067
HFRP-D9-b	92.3	0.0162	1,149	80.6	0.0129	1,040	1,330	20.6	0.0222	1,560	1,067
HFRP-D9-c	82.9	0.0223	1,548	80.6	0.0129	1,040	1,330				1,067

Specimen	Meas. E_{HF} [GPa]	Meas. $\varepsilon_{u\_C\_HF}$	Meas. $\sigma_{u\_C\_HF}$ [MPa]	$E_{HF}^{\dagger}$ [GPa]	$\varepsilon_{u\_C\_HF}^{\dagger}$	$\sigma_{u\_C\_HF}^{\dagger}$ [MPa]	$\sigma_{u\_C\_HF}^{\ddagger}$ [MPa]	Meas. E_{GF}^{*} [GPa]	Meas. $\varepsilon_{u\_G\_HF}$	Meas. $\sigma_{u\_G\_HF}$ [MPa]	$\sigma_{u\_G\_HF}^{\dagger}$ [MPa]
Average	89.4	0.0195	1,674	80.6	0.0129	1,040	1,330	65.3	0.023	1,763	1,067
HFRP-10-a	73.8	0.0209	1,544	79.4	0.0129	1,024	1,,317				1,078
HFRP-10-b	74.7	0.021	1,568	79.4	0.0129	1,024	1,317	100.8			1,078
HFRP-10-c	70.2	0.0205	1,439	79.4	0.0129	1,024	1,317				1,078
HFRP-D10-a	86.5	0.0187	1,621	79.4	0.0129	1,024	1,317				1,078
HFRP-D10-b	85	0.0171	1,396	79.4	0.0129	1,024	1,317	66.7	0.0205	1,494	1,078
HFRP-D10-c	88.3	0.0143	1,305	79.4	0.0129	1,024	1,317				1,078
Average	79.8	0.0188	1,479	79.4	0.0129	1,024	1,317	83.8	0.0205	1,494	1,078

HFRP = Hybrid FRP sheet; D = Ductile K type epoxy.

1 → (GF/CF = 1/1); 2 → (GF/CF = 2/1); 3 → (GF/CF = 3/1); 4 → (GF/CF = 4/1); 5 → (GF/CF = 5.1/1);
6 → (GF/CF = 6.1/1); 7 → (GF/CF = 7.1/1); 8 → (GF/CF = 8.1/1); 9 → (GF/CF = 9.1/1); 10 → (GF/CF = 10.1/1).

† = based on the rule of mixtures.
‡ = based on the rule of hybrid mixtures (Miwa and Horiba, 1994).
E_{HF} Elastic modulus of hybrid FRP sheet.
$\sigma_{u\_C\_HF}$ Stress at CF rupture of hybrid FRP sheet.
$\varepsilon_{u\_C\_HF}$ Strain at CF rupture of hybrid FRP sheet.
E_{GF}^{*} Average increase in stress of hybrid FRP sheet after CF rupture divided by increase in strain.
$\sigma_{u\_G\_HF}$ Stress at GF rupture of hybrid FRP sheet.
$\varepsilon_{u\_G\_HF}$ Strain at GF rupture of hybrid FRP sheet.
Conversion: 1 MPa = 0.145 ksi; 1 GPa = 145 ksi.

3. THE RULE OF MIXTURES

Chou and Kelly [3] and Manders and Bader [2] proposed a tensile stress model for hybrid carbon-glass FRP composites, as shown in Figure 3, based on the rule of mixtures. Points A and D denote the ultimate tensile stresses when GF and CF, respectively, are used alone (i.e., GF = 100%; CF = 100%). Also, the lines A-E and B-D represent the mean stresses in hybrid FRP when GF and CF fail, respectively. The CF with lower ε_{u_CF} than ε_{u_GF} (or higher $E_{\_CF}$ than $E_{\_GF}$) always fails prior to GF. To the right of Point C, after the first failure of CF, the hybrid FRP has a very low residual mean stress that is only provided by GF (i.e., brittle failure). To the left of Point C, even after the first failure of CF, the hybrid FRP with a relatively large amount of GF can sustain more loads without a drop in strength until the GF rupture. As such, pseudo-ductility can be achieved with this combination.

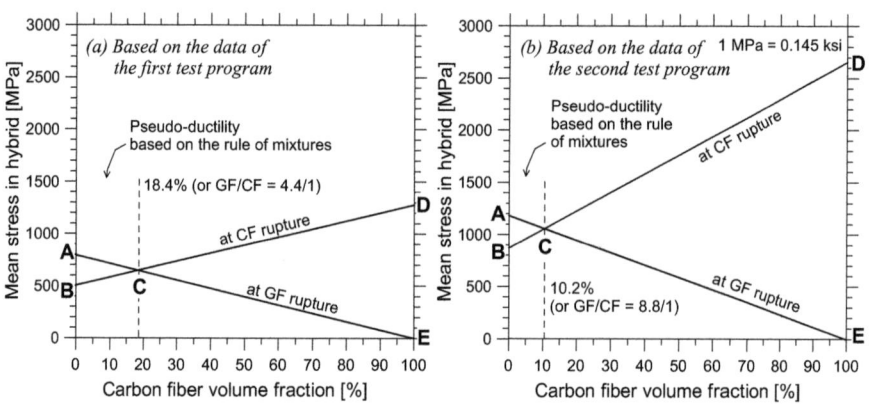

Figure 3 The Rule of Mixtures

Points A and D are taken as f_{u_CF} and f_{u_GF}, respectively, which can be obtained from the roving or sheet tests. The mean stress at Point B is calculated as ε_{u_CF} times E_{GF}, where ε_{u_CF} is the measured average ultimate strain of the CF roving/sheet and E_{GF} is the measured elastic modulus of the GF roving/sheet. Using the rule of mixtures and the material properties, the x-axis value at Point C is determined (e.g., (4.4/1) for Choi et al. [13]). The ratios of (4.4/1) and (8.8/1) are equivalent to the carbon volume fractions of 18.4%

14

and 10.2%, respectively. The y-axis value can be obtained from a cross point of two straight lines drawn in Figure 3(a) or 3(b).

To determine the stress (f_{HF}) in hybrid FRP for a given tensile strain (ε_{HF}), the rule of mixtures is applied as follows:

If $\varepsilon_{HF} \leq \varepsilon_{u\_CF}$ $$f_{HF} = \left[E_{CF}\left(\frac{A_{CF}}{A_{HF}}\right) + E_{GF}\left(\frac{A_{GF}}{A_{HF}}\right)\right]\varepsilon_{HF} \text{ or } f_{HF} = E_{HF}\varepsilon_{HF} \tag{3}$$

If $\varepsilon_{u\_CF} < \varepsilon_{HF} \leq \varepsilon_{u\_GF} \times (A_{GF}/A_{HF})$ $$f_{HF} = E_{GF}\varepsilon_{HF} \tag{4}$$

where ε_{HF} and f_{HF} are the strain and stress in hybrid FRP (variables), respectively; ε_{u_CF} and ε_{u_GF} are the ultimate strains of CF and GF, respectively; E_{CF} and E_{GF} are the elastic moduli of CF and GF, respectively; A_{CF} and A_{GF} are the cross-sectional areas of CF and GF in a hybrid sheet coupon, and A_{HF} is ($A_{CF} + A_{GF}$).

Equations (3) and (4) are consistent with Figure 3. For example, when the fiber roving properties reported by Choi et al. [13] are considered and the carbon volume fraction is 18.4%, the stress of f_{HF} is calculated to be [($0.184E_{CF} + 0.816E_{GF}$) × ε_{HF}] at CF rupture, which is represented by Point C in Figure 3(a). Once CF rupture occurs Equation (4) is applicable and the stress at GF rupture is assumed to be the same as [$\varepsilon_{u\_GF}E_{GF}$ × (A_{GF}/A_{HF})]. This failure is represented by the line A-E in Figure 3.

4. ASSESSMENT OF TEST RESULTS

Based on the information relevant to the mechanical properties of materials obtained in the test programs, an attempt was made to identify two different hybrid effects: 1) improved mechanical properties until the first peak, and 2) those after the first peak. Assessment of the data was carefully undertaken in this chapter.

4.1 FIRST TEST PROGRAM

First, the roving tensile properties were obtained (Table 1). Using each fiber's Specific Gravity (ρ_{CF} = 1.8; ρ_{GF} = 2.54) and measured weight per unit length, the cross-sectional area was determined to be 0.444 and 0.866 mm^2 (6.9 x 10^{-4} and 1.34 x 10^{-3} in^2) for the tested CF and GF rovings, respectively. The weight of the fiber roving was measured using a digital scale with an accuracy of +/–0.01 g (or 2.2 x 10^{-5} lbs).

Results from the tensile tests are presented in Table 1. In the remainder of the paper, the mean stress and strain (f_u and ε_u) from the non-impregnated (bare) coupon tests were used (though both results are similar). This is because epoxy resins are typically applied on only one side of the FRP sheet. The roving results for ultimate stress (f_u), ultimate strain (ε_u) and elastic modulus (E) in tension are substantially lower than the filament properties provided by the manufacturer (Table 1). This is due to the fragmentation process that generates unequal tension of filaments within a roving and failure strain variation between the filaments. Such phenomena are also seen from the behavior of conventional FRP sheets externally bonded to the concrete surface.

As indicated in Table 2 (compare Columns (4) and (7), and (11) and (12)), both stresses at CF and GF ruptures in hybrid FRP sheets with a (GF/CF) ratio of (8.8/1) were higher than the theoretical values, respectively. These positive hybrid effects can be seen as a result of the synergistic strengthening of both fibers. In particular, the ultimate stress and strain were increased by about 38% and 25%, respectively, compared with those of the GF roving. The corresponding stiffness after the first peak (CF rupture) was 54.5 MPa (7,903 ksi), about 20% larger than the elastic modulus (45 MPa or 6,525 ksi) of the GF roving. Tensile stress-strain relations of 2 samples of the hybrid FRP sheets are plotted in Figure

16

4(b), where the LVDT data were used to determine strains, in comparison with that of steel. Again, it is noted that the stress was calculated as the measured load divided by the cross-sectional areas of A_{HF} until the first peak (CF rupture) and A_{GF} after CF rupture.

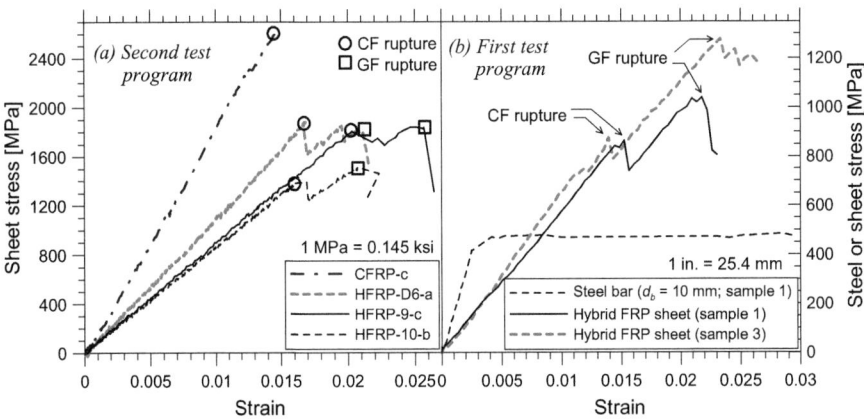

Figure 4 Measured Stress-Strain Curves of Selected Carbon FRP and Hybrid FRP Sheets and Conventional Steel

4.2 SECOND TEST PROGRAM

In this study, the roving properties of each fiber were not obtained; rather, the impregnated carbon FRP and glass FRP sheets were tested in tension to obtain each fiber's properties. Unfortunately, the results of the impregnated glass FRP sheets were misplaced; thus, the ultimate strain (ε_{u_GF} = 0.0176) from the roving tests of Choi et al. [13] is adopted for further analysis, as the fibers used for both programs were manufactured by the same vendor. For the elastic modulus of GF, the average stiffness of the hybrid sheets after CF rupture can be generally taken into account assuming that the Young's modulus (E_{GF} = 67.3 GPa; 9,760 ksi) is almost the same as that of GF. Finally, the ultimate stress (f_{u_GF} = 1,185 MPa; 172 ksi) is taken as the product of 0.0176 and 67.3 GPa (9,760 ksi).

Figure 4(a) shows selected results of the impregnated carbon FRP and hybrid FRP sheets. The stress was taken to be the load divided by the cross-sectional area of the whole fibers until CF rupture and by the area of the glass fibers only after CF rupture. The cross-sectional area of the sheet was determined based on the number of each fiber roving and its cross-sectional area estimated using the roving weight and Specific Gravity (ρ_{CF} = 1.8; ρ_{GF} = 2.54), where the weight was measured using a micro-digital scale with an accuracy of +/–1 x 10^{-5} g (or 2.2 x 10^{-8} lbs). This does not include the area of the epoxy. In addition, it appears that the type of epoxy resins did not affect the tensile behavior of the sheets much. There was no clear evidence of different performance between the sheets impregnated using the J type and K type epoxy resins.

The ultimate stresses (average f_{u_CF} = 2,656 MPa or 372 ksi) and elastic moduli (average E_{CF} = 202 MPa or 29,360 ksi) of the impregnated carbon FRP sheets are lower than those (f_{u_CF} = 4,900 MPa or 710.5 ksi; E_{CF} = 230 GPa or 33,350 ksi) of a carbon filament that the manufacturer reported. The differences are much less compared with those between the filament and rovings. The average ultimate strain (ε_{u_CF}) of the CF sheets is 0.013, which is smaller than all but three of 91 ultimate strains (average $\varepsilon_{u\_C\_HF}$ = 0.018) of the hybrid sheet coupons at CF rupture. This is evidently due to the synergistic hybrid effect. For the hybrid FRP sheets, the ratio of (GF/CF) also affected the overall behavior. As the (GF/CF) ratio increased, both the strains at CF and GF ruptures generally increased (Table 3). In the following chapter, more detailed investigations are conducted in connection with the theoretical models.

5. COMPARISON WITH THE RULE OF MIXTURES

Experimental results corresponding to line B-D in Figure 3 are obtained using the measured strains of hybrid FRP coupons at CF rupture and the measured loads, summarized in Column (4) of Tables 2 and 3, and depicted in Figures 5(a) and 5(b). The significantly increased stresses relative to line B-D are noted. The average strain ($\varepsilon_{u\_C\_HF}$) at CF rupture is also about 35% higher than ε_{u_CF}. For (GF/CF) ratios higher than (4.4/1), an increase of about 45% occurred, whereas for (GF/CF) ratios lower than (4.4/1), the increase was about 22%. In general, the value of $\varepsilon_{u\_C\_HF}$ increases as the (GF/CF) ratio increases. This indicates that a constant increase in failure strain, when the hybrid effect is expected, may not be valid (e.g., 50% or 0.01 strain increase). The difference between the measured stresses of $f_{u\_C\_HF}$ and the predicted stresses of $f_{u\_C\_HF}$ based on Equation (2) is only about 10% (see Columns (4) and (8) of Tables 2 and 3), indicating that the rule of "hybrid" mixtures suggested by Miwa and Horiba [4] works better than the rule of mixtures for fiber composites, which underestimates the experimental values by about 35%.

Based on these results, the positive hybrid effect is evident for hybrid FRP sheets that are usually used for repair and retrofit of the concrete structures, and these hybrid effects include the stress and strain of the hybrid FRP sheet at CF rupture (but not Young's modulus). The Young's modulus of the hybrid FRP sheet until CF rupture is almost the same as that of the carbon FRP sheet (1% difference on average).

Similarly, experimental results corresponding to the line A-E in Figure 3 are examined. The average stress ($f_{u\_G\_HF}$ = 1,094 MPa or 158.6 ksi) at GF rupture was monitored from 3 specimens of Choi et al. [13] with the (GF/CF) ratio of (8.8/1). This is about 40% higher than the corresponding point [$f_{u\_GF} \times (A_{GF}/A_{HF})$] = 771 MPa or 103.2 ksi) in Figure 5(c) (see Column (12) of Table 2). The average ultimate strain ($\varepsilon_{u\_G\_HF}$ = 0.022) of the hybrid FRP is also higher by 40% relative to the product (= 0.0158) of ε_{u_GF} and (A_{GF}/A_{HF}). This is clearly inconsistent with the rule of mixtures. The average increase in stress after CF rupture divided by the average increase in strain (54.7 GPa or 7,076 ksi) is also somewhat higher than the elastic modulus (E_{GF} = 45 GPa or 6,525 ksi) of the GF roving (by about 20%).

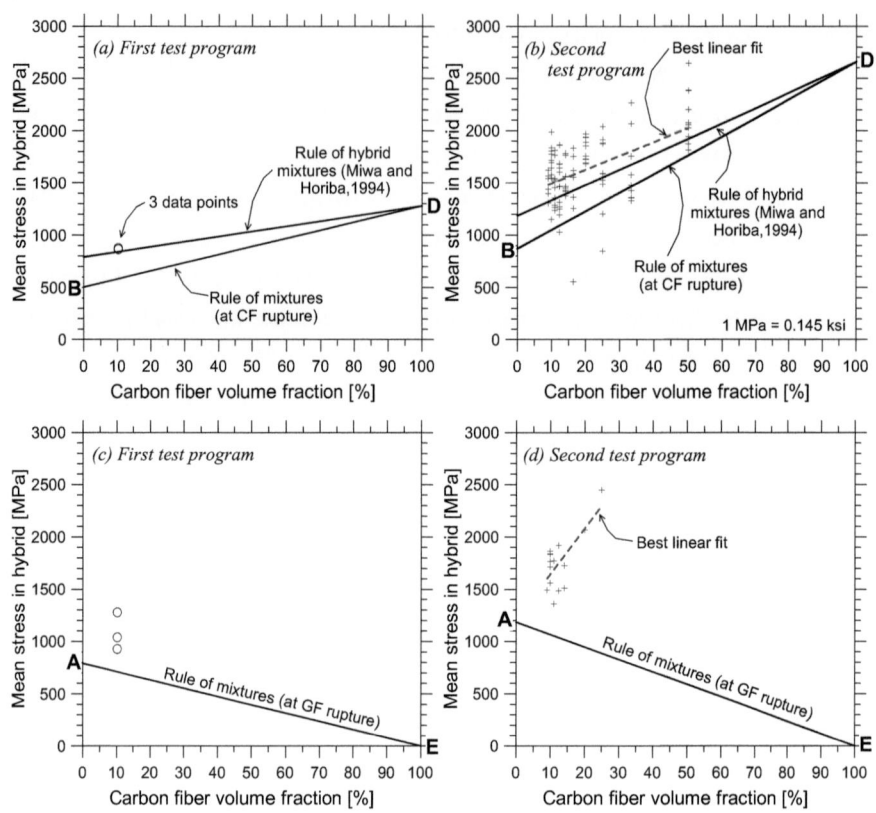

Figure 5 Identification of Hybrid Effects

Unlike the test program of Choi et al. [13], the GF rupture was not captured well for many specimens of the second test program because of the strain gauge or epoxy failure prior to GF rupture, and the disruption of testing once the strain gauge or epoxy failed. Therefore, direct comparisons between $f_{u\_G\_HF}$ and (f_{u_GF} x [GF/(GF+CF)]) or between ε_{u_GF} and ($\varepsilon_{u\_G\_HF}$ x [GF/(GF+CF)]) are not possible due to the absence of the GF properties. The averages of the 14 measured values of $f_{u\_G\_HF}$ and $\varepsilon_{u\_G\_HF}$ are 1,759 MPa (255 ksi) and 0.022, respectively, which are generally higher than the roving test results of Choi et al. [13]. Interestingly, there is a tendency of increasing failure stress ($f_{u\_G\_HF}$) with decreasing

(GF/CF) ratio or increasing carbon volume fraction (Figure 5). This means that both the initial stiffness and pseudo-ductility can be obtained even with lower (GF/CF) ratios (see Figures 4(a) and 5). While the trend opposite to the rule of mixtures is noteworthy, the data appear not to indicate different ultimate strains depending on the (GF/CF) ratio (versus ultimate stresses). There is a need for further experimental investigation for the ultimate stress and strain in relation with the (GF/CF) ratio.

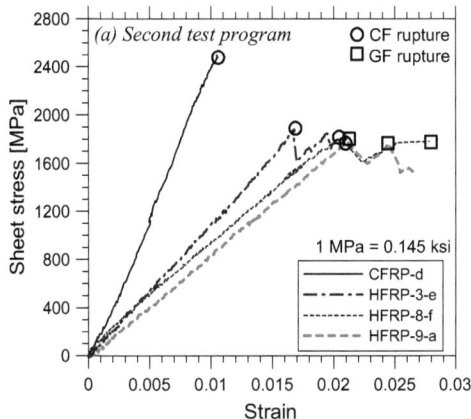

Figure 6 Pseudo-Ductility of Hybrid FRP Sheets Noted From Measured Stress-Strain Curves

Based on the two independent programs investigated, it can be concluded that the hybrid effects are positive in terms of the ultimate stress and strain at GF rupture for all (GF/CF) ratios. Note that these positive effects are contrary to the prediction made by Pan and Postle [8] for micro-fibers embedded in the matrix. This seems to be related to the degree of coupling between two different fibers. The use of hybrid FRP sheets consisting of CF and GF rovings has been proven to be a very effective means to promote synergistic hybrid effects. In a similar manner, the hybrid effect of uniaxial hybrid FRP sheets that are made of 3 different fiber rovings could be investigated.

Figure 4(b) shows the stress-strain relationship for the 2 coupons tested by Choi et al. [13], where pseudo-ductility was observed. This is consistent with the rule of mixtures which suggests the recovery of the stress after CF rupture when a point representing the (GF/CF) ratio is located to the left of Point C in Figure 3. On the other hand, the data reported from the second test program exhibit a high degree of ductility after CF rupture even for low (GF/CF) ratios (Figures 4(a) and 6). Because the data are quite limited, additional tests would be helpful to evaluate the stress recovery after CF rupture.

6. PROPOSED STRESS-STRAIN RELATIONSHIP FOR HYBRID FRP SHEETS

When the moment and shear capacities of concrete members strengthened with hybrid FRP sheets are determined, a stress-strain or force-strain relationship of the hybrid FRP sheets would be needed. The stress-strain relationship of Equations (3) and (4), which is based on the rule of mixtures, does not account for the identified positive hybrid effects. In this study, based on the review on the test results, the following stress-strain relationship for hybrid carbon-glass hybrid FRP sheets is proposed.

If $\varepsilon_{HF} \le \varepsilon_{u\_C\_HF}$ $\qquad$ $f_{HF} = E_{HF}\varepsilon_{HF} = \left[E_{CF}\left(\frac{A_{CF}}{A_{HF}}\right) + E_{GF}\left(\frac{A_{GF}}{A_{HF}}\right) \right] \varepsilon_{HF}$ $\qquad$ (5)

If $\varepsilon_{u\_C\_HF} < \varepsilon_{HF} \le \varepsilon_{u\_G\_HF}$ $\qquad\qquad$ $f_{HF} = E_{GF}\varepsilon_{HF}$ $\qquad\qquad$ (6)

where $\varepsilon_{u\_C\_HF}$ is taken as $(f_{u\_C\_HF}/E_{HF})$; $f_{u\_C\_HF}$ can be estimated using Eq (2); and $\varepsilon_{u\_G\_HF}$ is the strain of hybrid FRP sheets at GF rupture, suggested to be approximately 0.022 when a (GF/CF) ratio is larger than (4/1), respectively. The suggested values are based on the re-assessed data from the two test programs. It is noted that the elastic moduli in the proposed stress-strain relationship do not reflect the synergistic positive hybrid effect since the increased properties are minor. Also, note that after CF rupture, the stress of f_{HF} should be paired with A_{GF} (not A_{HF}) to determine the force, assuming that CF no longer resists any tension.

The analytical stress-strain relationships with and without consideration of hybrid effects are depicted in Figure 7 for a variety of (GF/CF) ratios. The former is expressed in the form of Equation (5) and (6), while the latter is expressed in the form of Equations (3) and (4). It is noted that $\varepsilon_{u\_C\_HF}$ is a function of the (GF/CF) ratio, whereas a fixed value of $\varepsilon_{u\_G\_HF}$ of 0.022 is proposed. Also, strains of ε_{u_CF} and ε_{u_GF} are suggested to be 0.013 and 0.018, respectively, and elastic moduli of E_{CF} and E_{GF} to be 202 GPa (29,300 ksi) and 67 GPa (9,700 ksi). If the rule of mixtures is applied, pseudo-ductility can only be achieved when a

(GF/CF) ratio is not less than (8.8/1) (Figure 7(a)). If the hybrid effect is considered, which has been demonstrated in this study, pseudo-ductility is characterized in almost all (GF/CF) ratios (Figure 7(b)). Furthermore, a strain at CF rupture is even higher than that at the second fiber (GF) rupture without the consideration of the hybrid effect (Figures 7(c) and 7(d)). This is one of the greatest advantages that hybridized carbon and glass fibers can offer. Regarding the effect of the (GF/CF) ratio, the hybrid model shows better performance in terms of the stiffness for lower (GF/CF) ratios, and the same ductility for all (GF/CF) ratios, still substantially higher than that of CF or GF. Such behaviors are also seen in Figures 4(a) and 6. However, a high (GF/CF) ratio (e.g., 8.7/1) is recommended because the reduction of the cross-sectional area of the FRP sheet with a very low (GF/CF) is drastic and cost effectiveness is at its maximum when the highest (GF/CF) ratio is used.

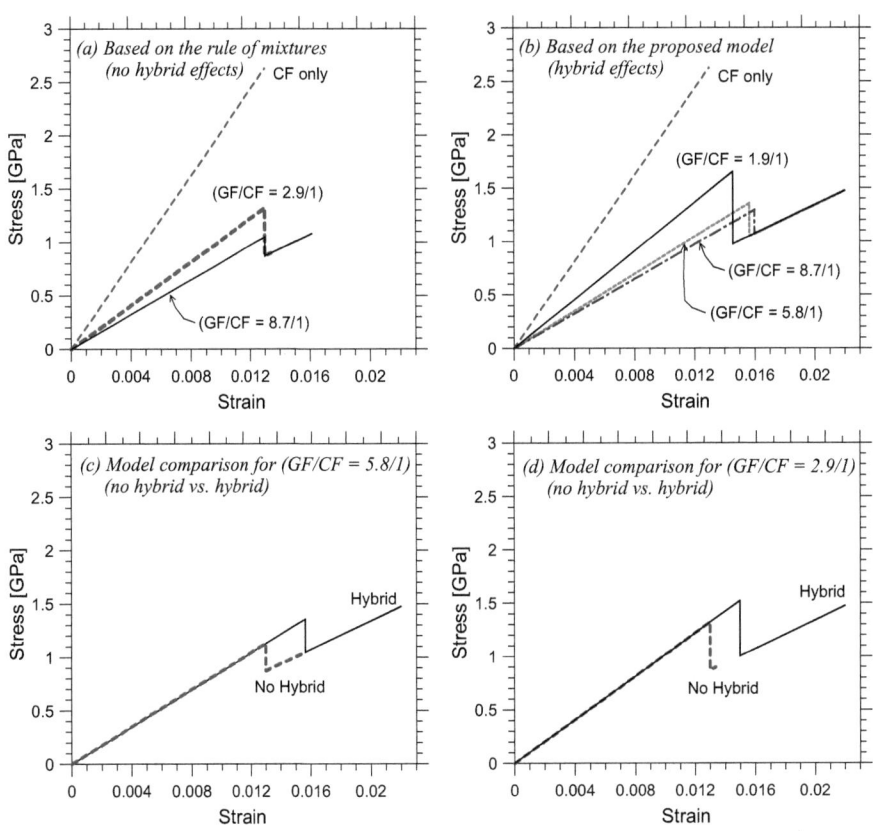

Figure 7 Models for Stress-Strain Relationship of Hybrid FRP Sheets with Various (GF/CF) Ratios

7. CONCLUSION

The purposes of the use of uniaxial hybrid FRP sheets in the repair of existing concrete structures are to achieve pseudo-ductility and utilize their synergistic hybrid effects. In this study, the tensile test results from a total of 94 hybrid carbon-glass FRP sheets and 47 carbon and glass fiber rovings or sheets were evaluated and reviewed in depth. Based on this review, a number of conclusions are made as follows:

1. All three types of grips developed by Choi et al. (2009) [13] for the roving tests are effective. In particular, the grip with 90° sandwich laminates using the same fiber rovings appears to be very sound.

2. The two epoxy resins (J and K types) sustained strains up to about 0.02 and 0.03, respectively; however, an ultimate strain of about 0.04 is recommended to prevent epoxy failure prior to fiber rupture. Neither type of the epoxy resins affected the tensile behavior of the sheets much.

3. The elastic moduli of hybrid FRP sheets generally correspond to the rule of mixtures.

4. The strains at CF and GF ruptures of the hybrid sheets are about 0.018 and 0.022, respectively, on average, which are substantially higher than the ultimate strains of each CF and GF (0.013 and 0.018). A trend of increased strain at CF rupture for increased (GF/CF) ratio was observed, while there is no clear indication of different strains at GF rupture depending on the (GF/CF) ratio.

5. The stresses at CF rupture of the hybrid sheets are also significantly higher than those predicted based on the rule of mixtures, differing by about 40%, but are quite close to those predicted based on the rule of hybrid mixtures developed by Miwa and Horiba (1994) (within about 10% difference). Also, the stresses at GF rupture are considerably higher than those predicted based on the rule of mixtures, by about 80%. The discovery of these positive hybrid effects is a significant advance.

6. A general trend of increased stress at GF rupture for decreased (GF/CF) ratio was observed. This signals that both the initial stiffness and pseudo-ductility could be obtained even with a very low (GF/CF) ratio. However, given the limited data, additional research would be needed to verify this trend.

7. The identified synergistic hybrid effects are evident for all (GF/CF) ratios. When carbon and glass fibers are hybridized, an ultimate strain at the first fiber (carbon) rupture could be even higher than the ultimate strain of the second fiber (glass only).

8. The synergistic positive hybrid effects at both CF and GF ruptures might be shown only in the hybrid FRP sheets that are made of CF and GF rovings. If each fiber had been mixed in a roving, the positive hybrid effects might have not been found.

REFERENCES

1. Phillips, L. N., "The Hybrid Effect-Does It Exist?," *Composites*, 7, Jan. 1976, pp. 7-8.

2. Manders, P. W. and Bader, M. G., "The Strength of Hybrid Glass/Carbon Fibre Composites: Part 1 – Failure Strain Enhancement and Failure Mode," *Journal of Materials Science*, 16, 1981, pp. 2233-2245.

3. Chou, T.-W. and Kelly, A., "Mechanical Properties of Composites," *Annual Reviews of Material Science*, 10, 1980, pp. 229-259.

4. Miwa, M. and Horiba, N., "Effects of Fibre Length on Tensile Strength of Carbon/Glass Fibre Hybrid Composites," *Journal of Materials Science*, 29, 1994, pp. 973-997.

5. Bunsell, A. R. and Harris, B., "Hybrid Carbon and Glass Fibre Composites," *Composites*, 5, 1974, pp. 157-164.

6. Aveston, J. and Sillwood, J. M., "Synergistic Fibre Strengthening in Hybrid Composites," *Journal of Materials Science*, 11(10), 1976, pp. 1877-1883.

7. Marom, G., Fischer, S., Tuler, F. R. and Wagner, H. D., "Hybrid Effects in Composites: Conditions for Positive or Negative Effects versus Rule of Mixtures," *Journal of Materials Science*, 13, 1978, pp. 1419-1426.

8. Pan, N. and Postle, R., "Tensile Strengths and the Hybrid Effects of Hybrid Fibre Composites: A Probabilistic Approach," *Transitions of the Royal Society in London*, Series A., 354, 1996, pp. 1875-1897.

9. Harris, H. G.; Somboonsong, W.; and Frank, K. K., "New Ductile Hybrid FRP Reinforcing Bar for Concrete Structures," *ASCE Journal of Composites for Construction*, 2(4), 1998, pp. 28-36.

10. Grace, N. F.; Abel-Sayed, G.; and Ragheb, W. F., "Strengthening of Concrete Beams Using Innovative Ductile Fiber-Reinforced Polymer Fabric," *ACI Structural Journal*, 99(5), Sept.-Oct. 2002, pp. 692-700.

11. Nanni, A., Henneke, M. J. and Okamoto, T., "Tensile Properties of Hybrid Rods for Concrete Reinforcement," *Construction and Building Materials*, 8(1), 1994, pp. 27-34.

12. Cox, H. L., "The Elasticity and Strength of Paper and Other Fibrous Materials," *British Journal of Applied Physics*, 3, 1952, pp. 72-79.

13. Choi, D.-U., Kang, T. H.-K., Ha, S.-S., Kim, K.-H. and Kim, W., "Flexural and Bond Behavior of Concrete Beams Strengthened with Hybrid Carbon-Glass Fiber-Reinforced Polymer Sheets," *ACI Structural Journal*, 108(1), Jan. 2011.

14. Song, H., Min, C. and Lee, C., "Tensile Behavior of Hybrid Fiber," *Academic Conference Proceedings of Korean Society of Civil Engineers*, Gangwon-do, Korea, 2009, 4 pp. (in Korean)

15. Kang, T. H.-K., Kim, W., Hufnagel, A., Ary Ibrahim, M., Huang, Y., Choi, D.-U., Lee, C. Y., and Holliday, L., "Repair and Retrofit of Concrete Girders Using Hybrid FRP Sheets," Final Report, OTCREOS10.1-21-F, Oklahoma Transportation Center, Midwest City, Okla., 2012, 83 pp.

16. ACI Committee 440, "Guide for the Design and Construction of Externally Bonded FRP Systems for Strengthening Concrete Structures (ACI 440.2R-02)," ACI, Farmington Hills, Mich., 2002, 45 pp.

17. ASTM International, "ASTM D3039/D3039M-08 Standard Test Method for Tensile Properties of Polymer Matrix Composite Materials," ASTM International, West Conshohocken, Penn., 2008, 13 pp.

18. CAN/CSA S806-02 (R2007), "Design and Construction of Building Components with Fibre-Reinforced Polymers," CSA, Mississauga, Canada, 2007, pp. 206.

19. ASTM International, "ASTM D638-08 Standard Test Method for Tensile Properties of Plastics," ASTM International, West Conshohocken, Penn., 2008, 16 pp.

20. ACI Committee 503, "Use of Epoxy Compounds with Concrete (ACI 503R-93; Reapproved 1998)," ACI, Farmington Hills, Mich., 1998, 28 pp.

APPENDIX – *Tensile Test results*

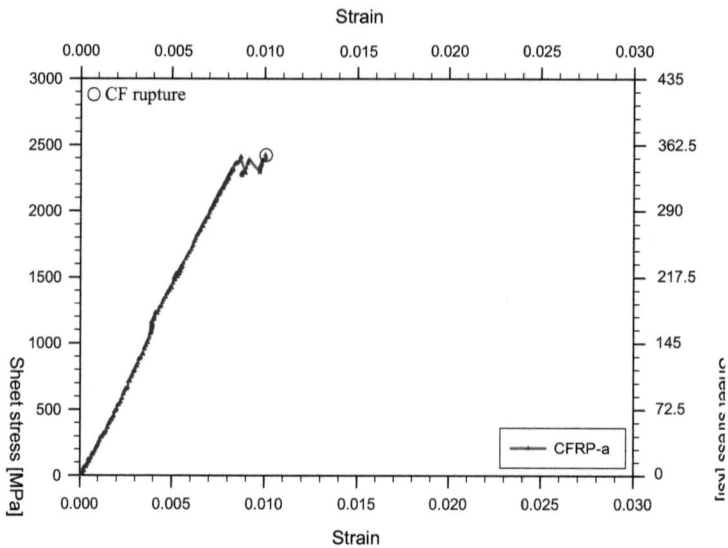

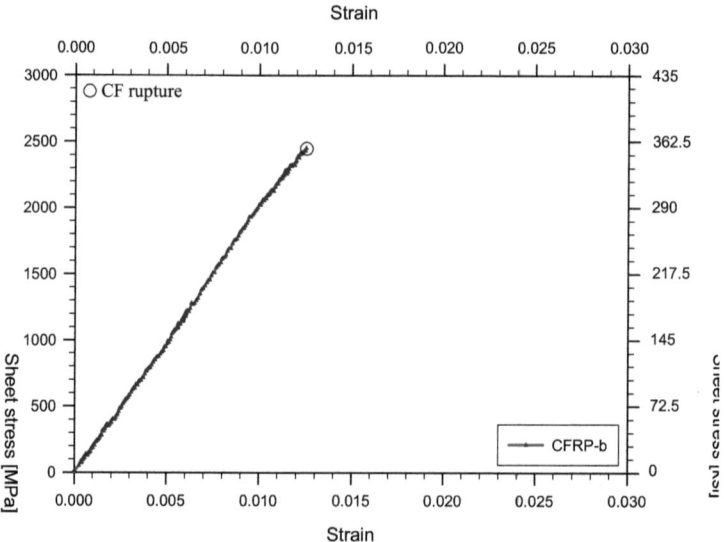

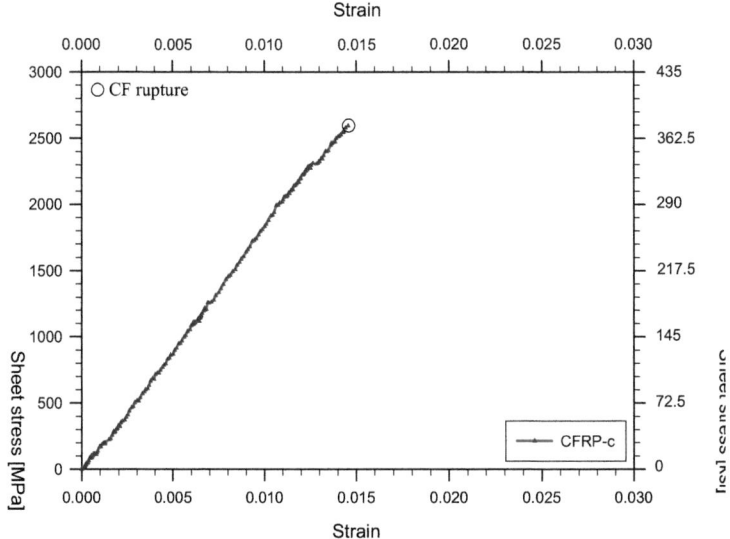

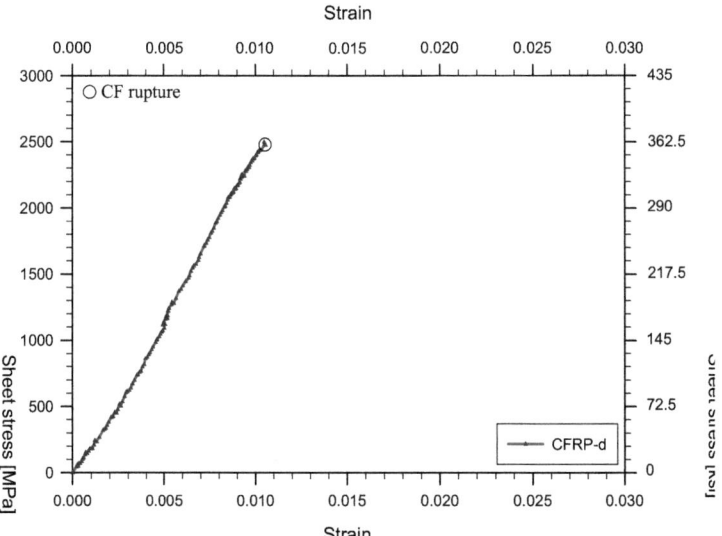

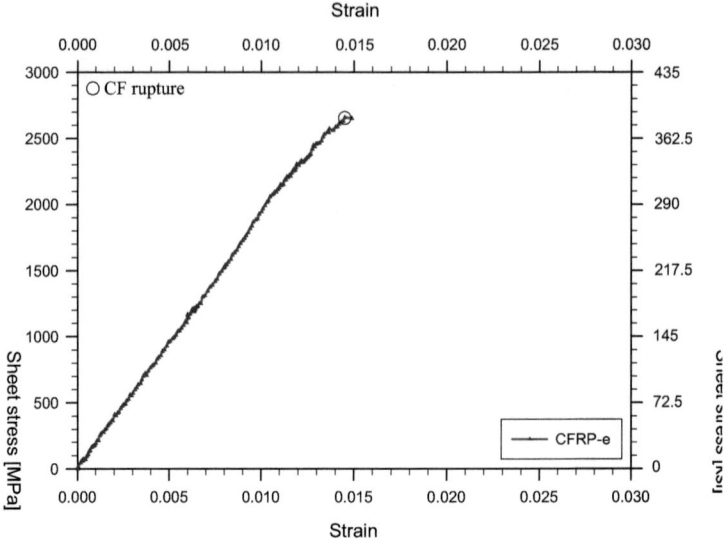

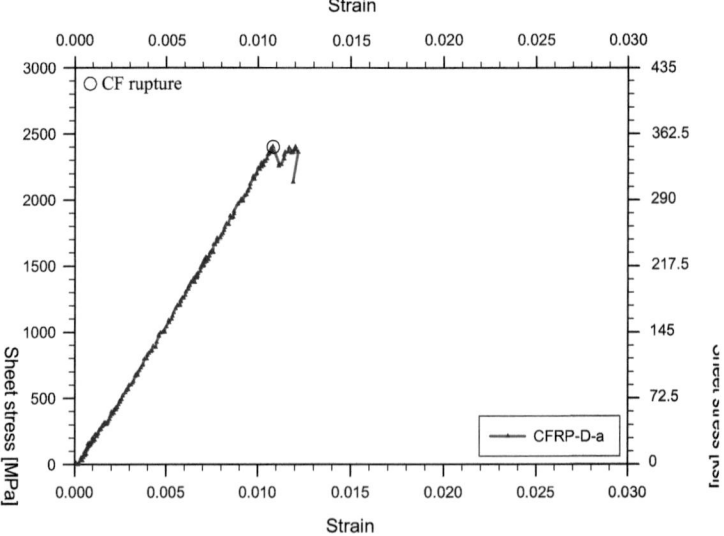

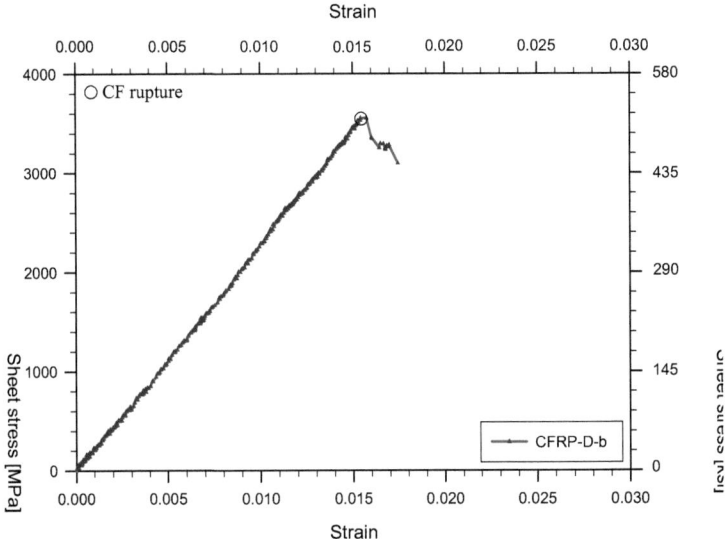

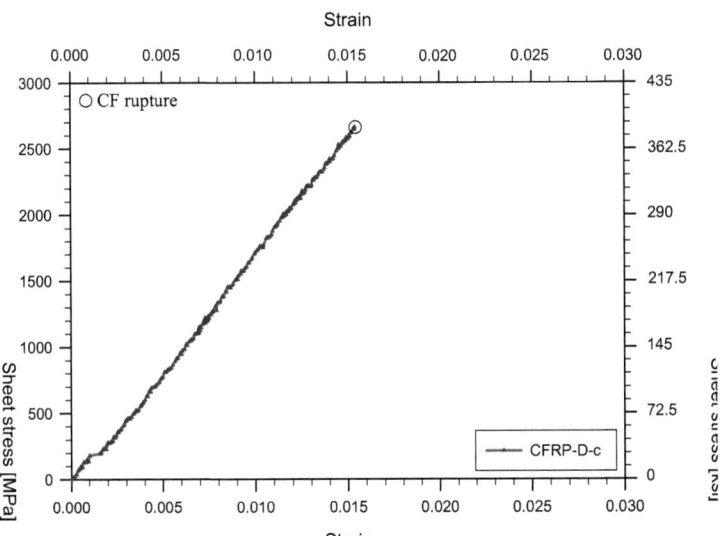

33

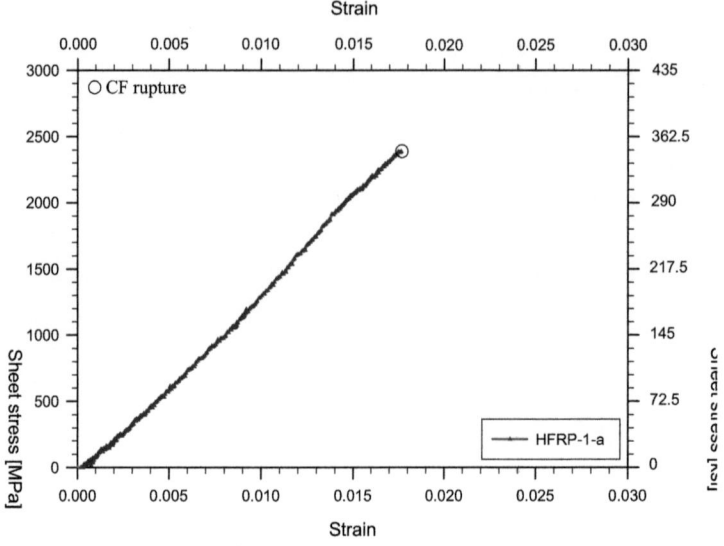

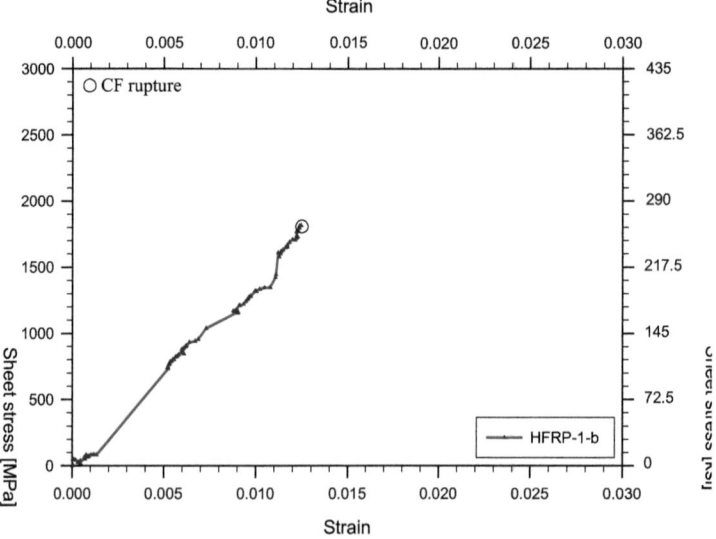

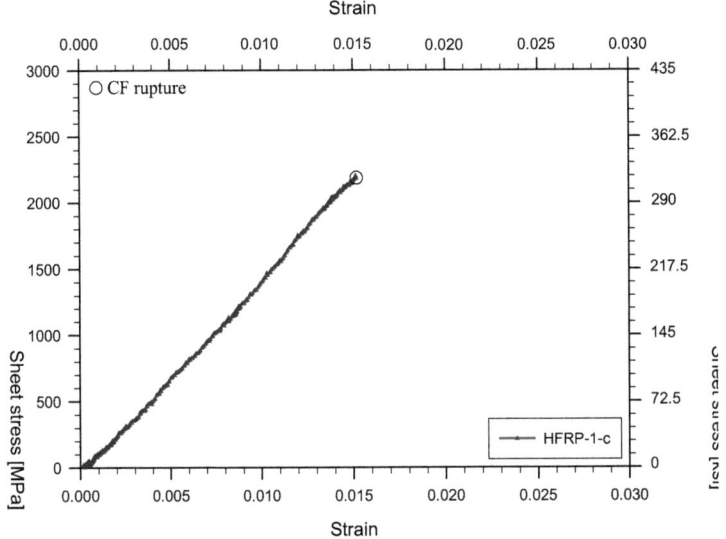

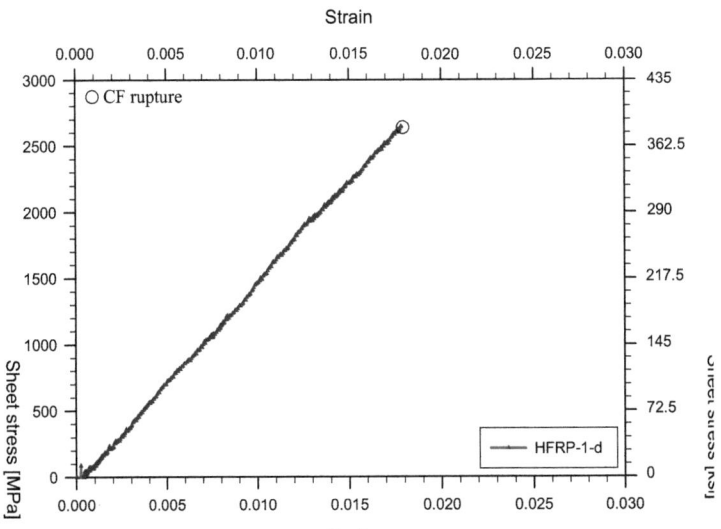

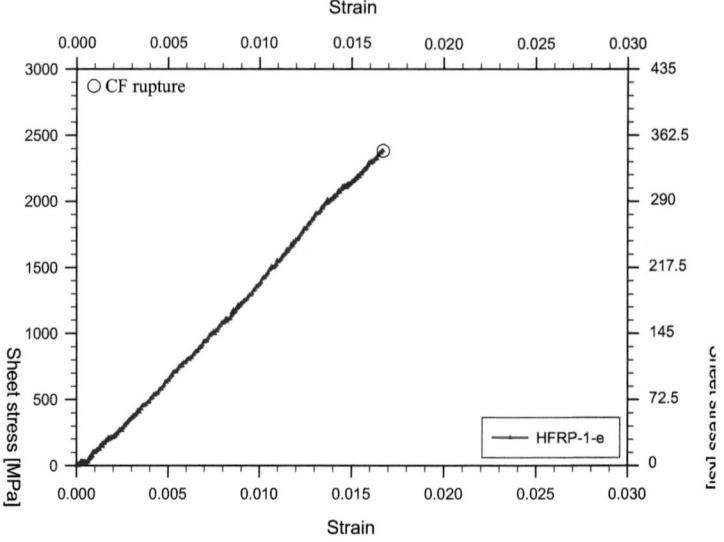

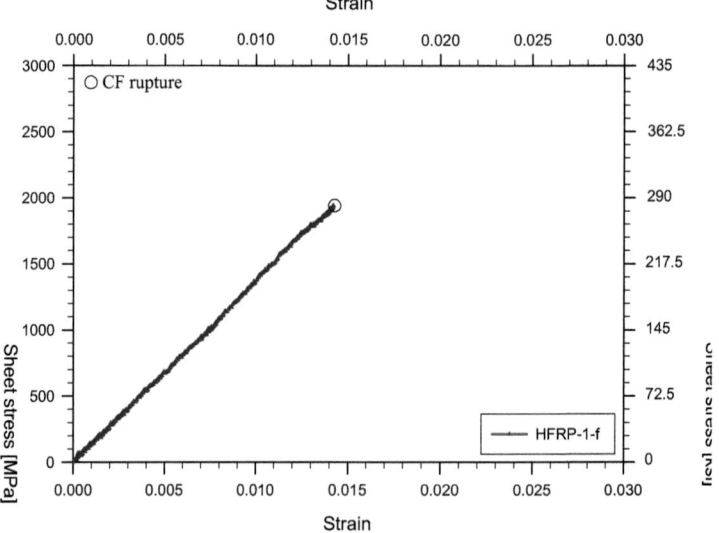

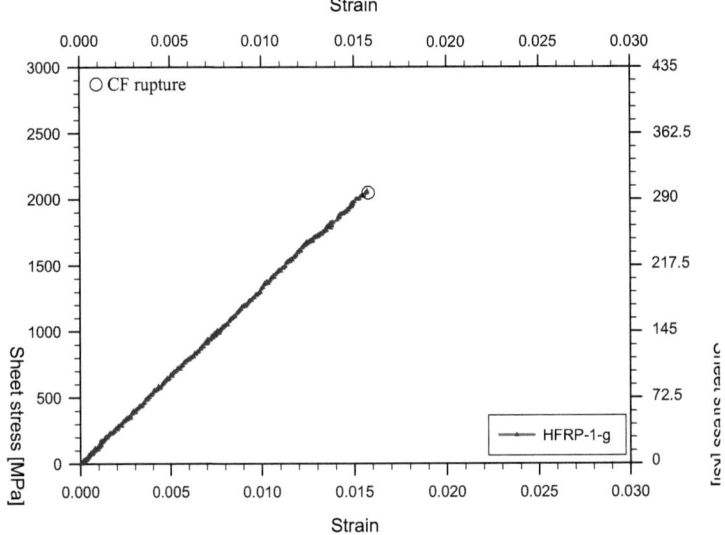

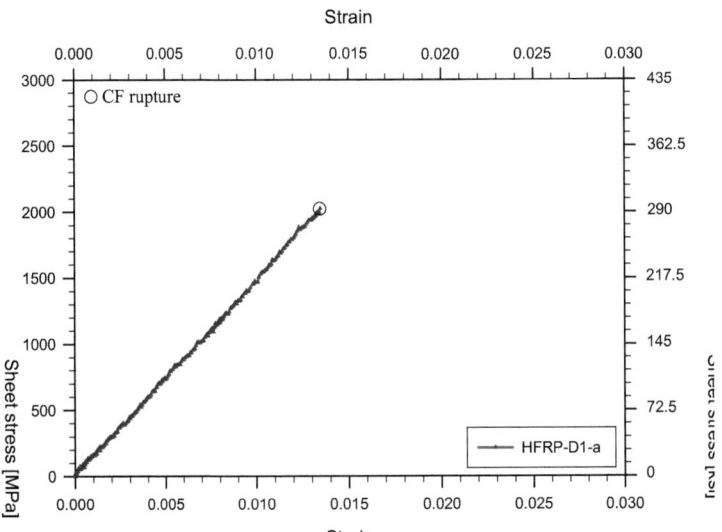

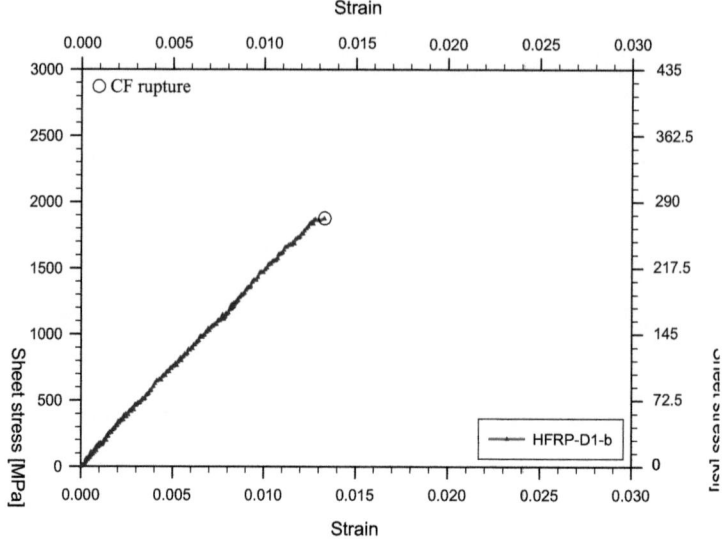

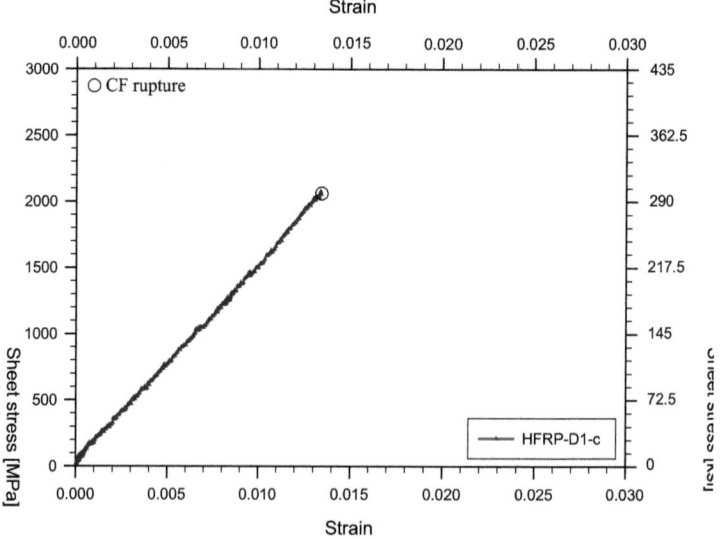

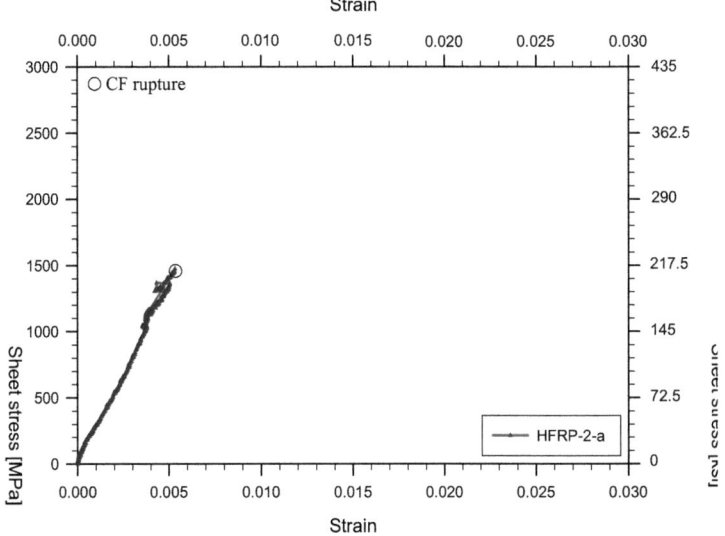

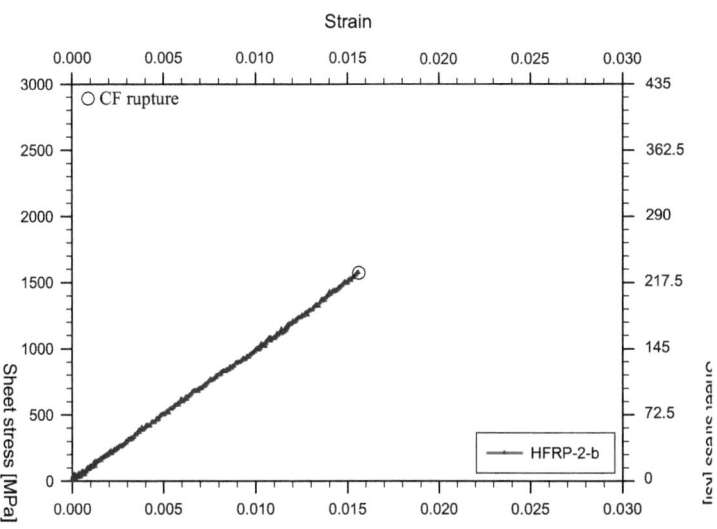

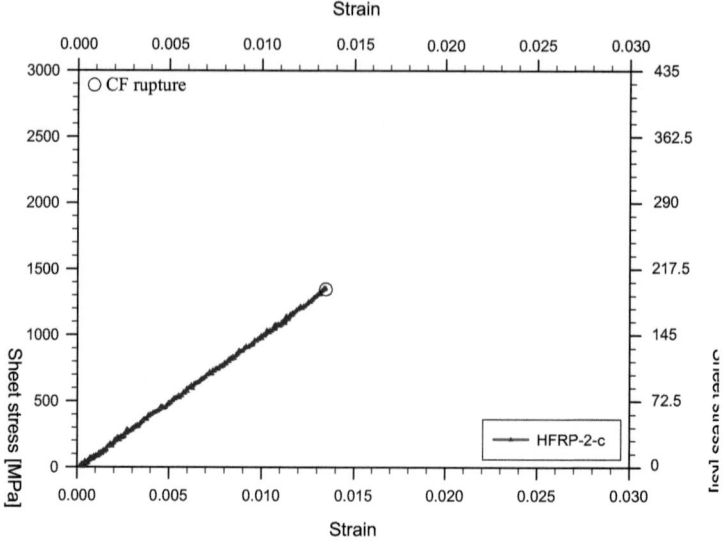

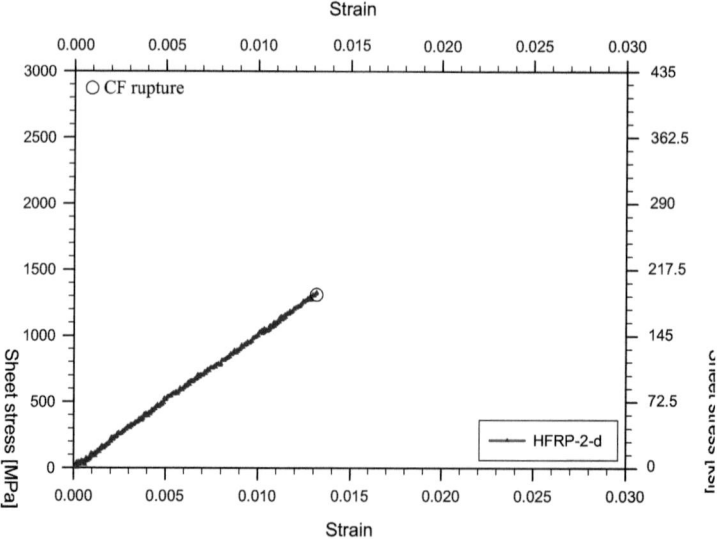

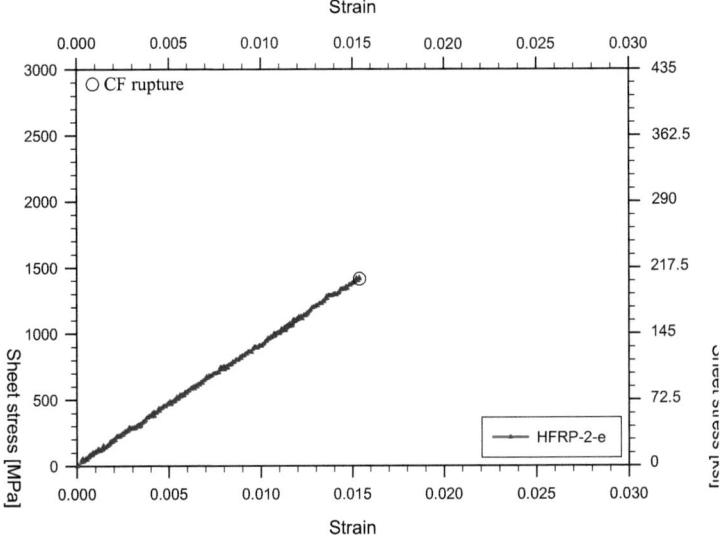

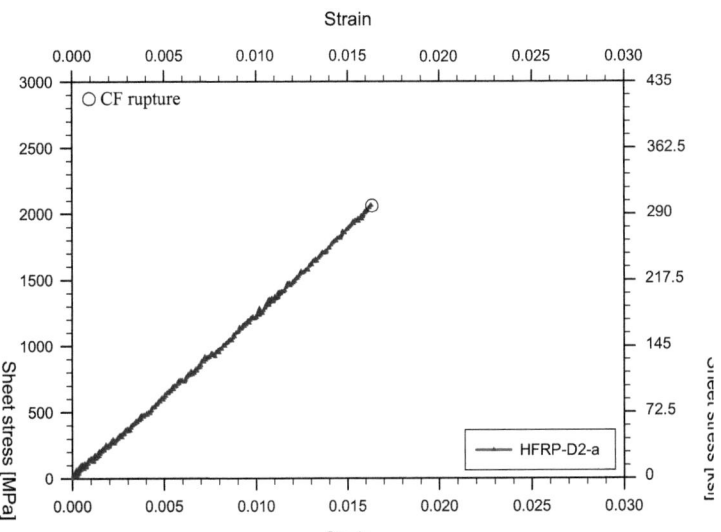

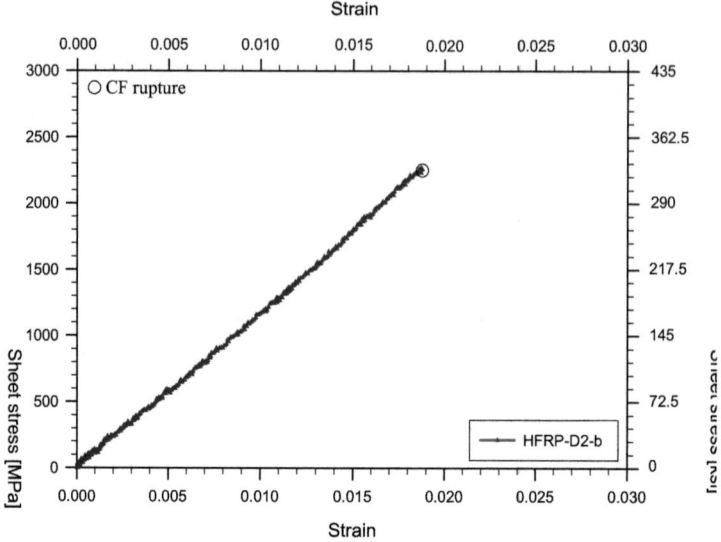

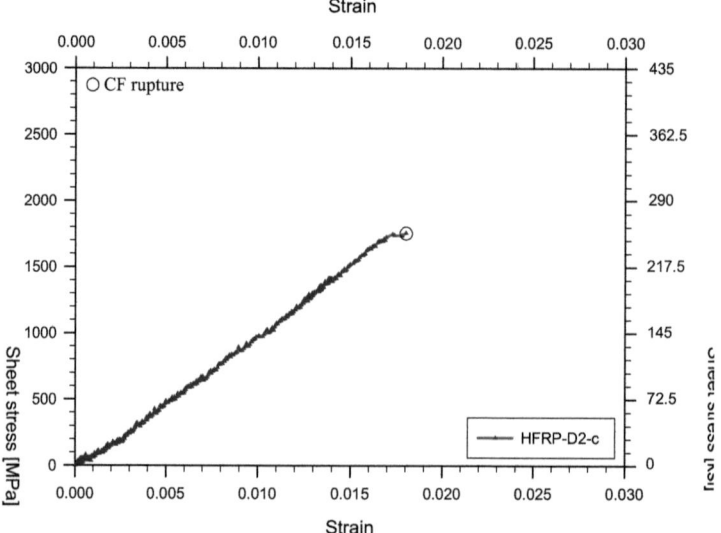

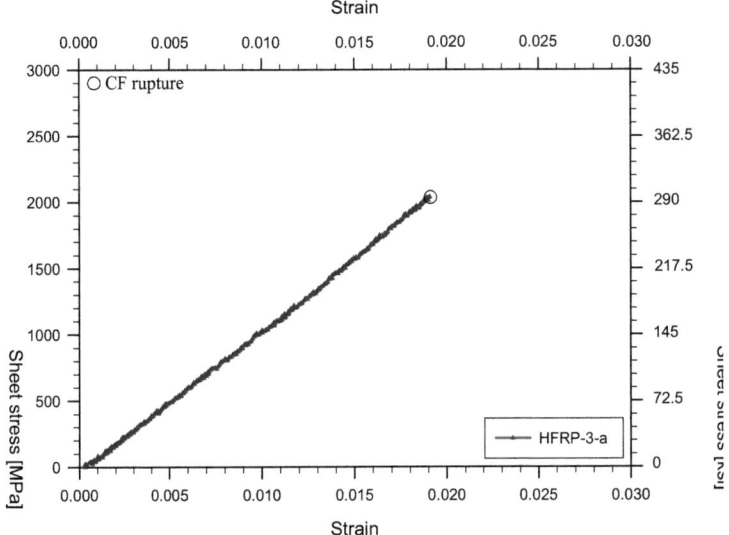

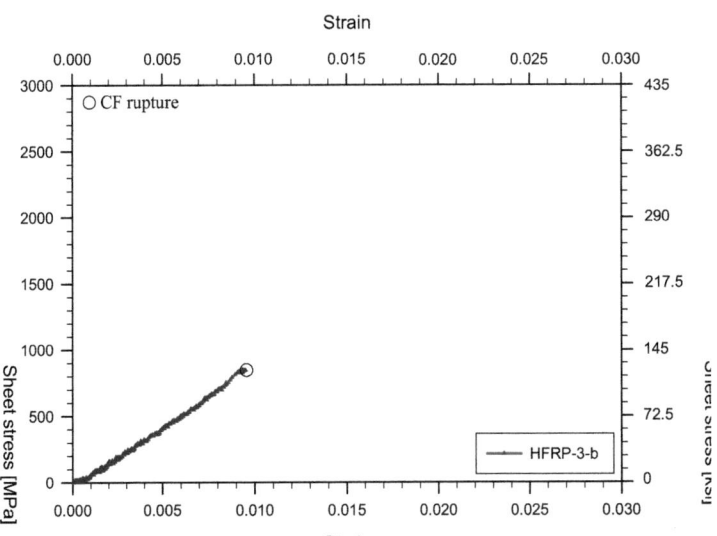

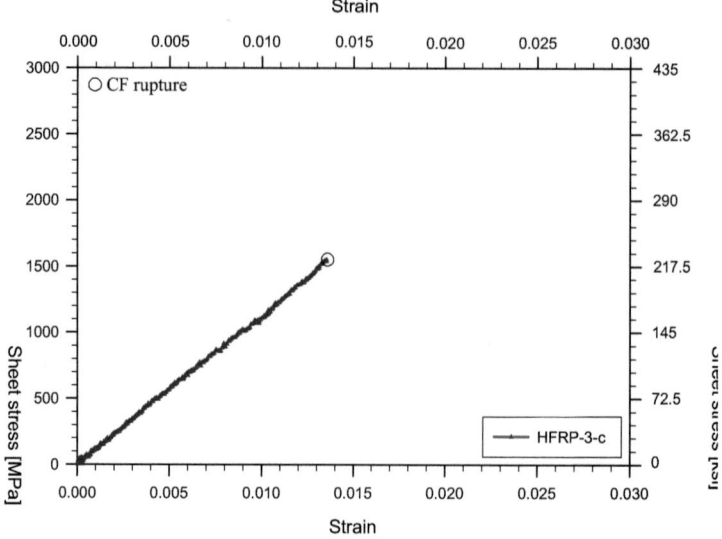

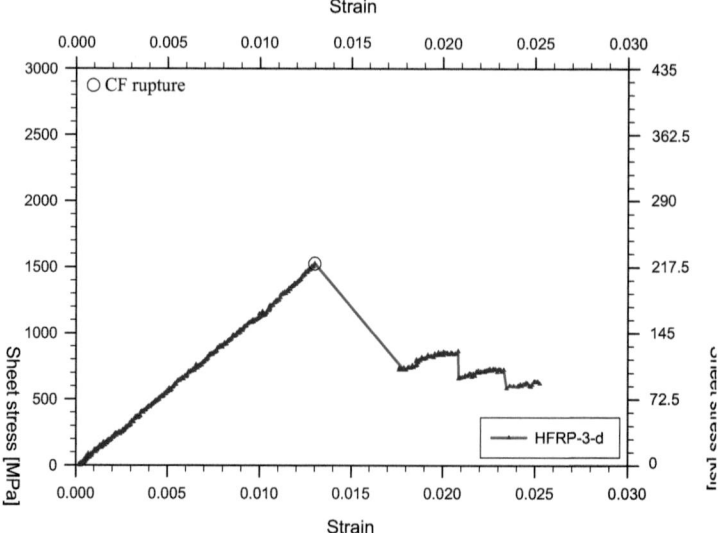

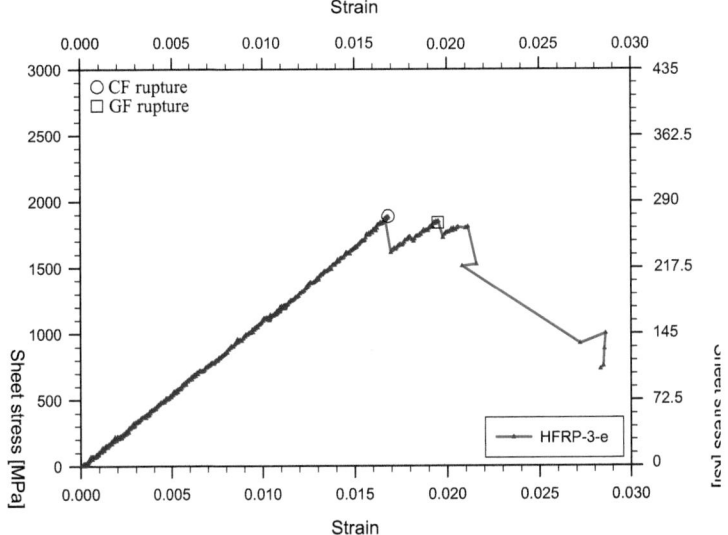

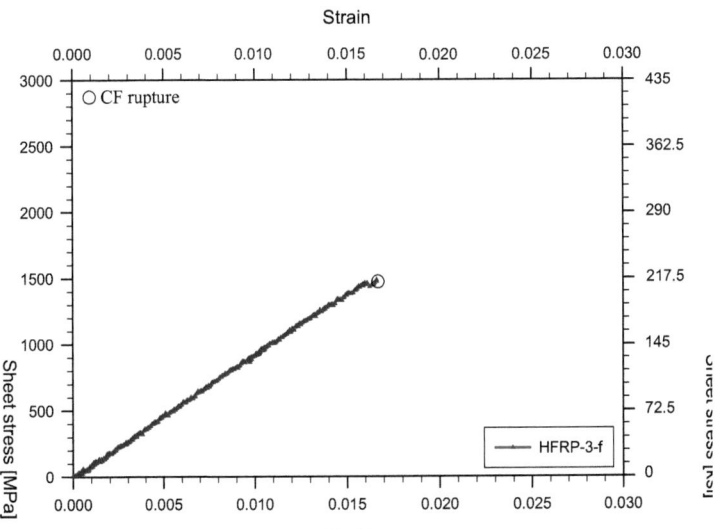

45

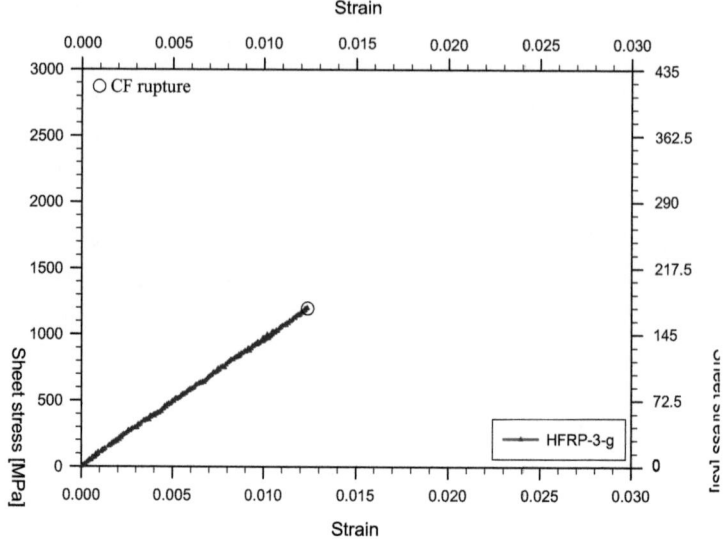

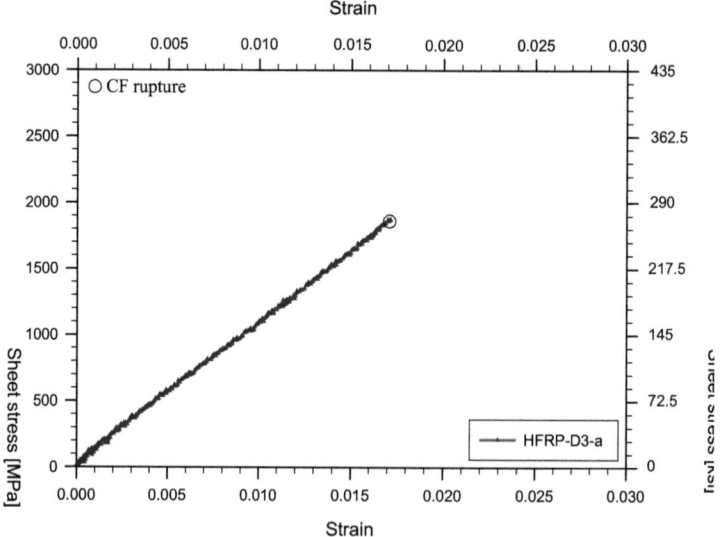

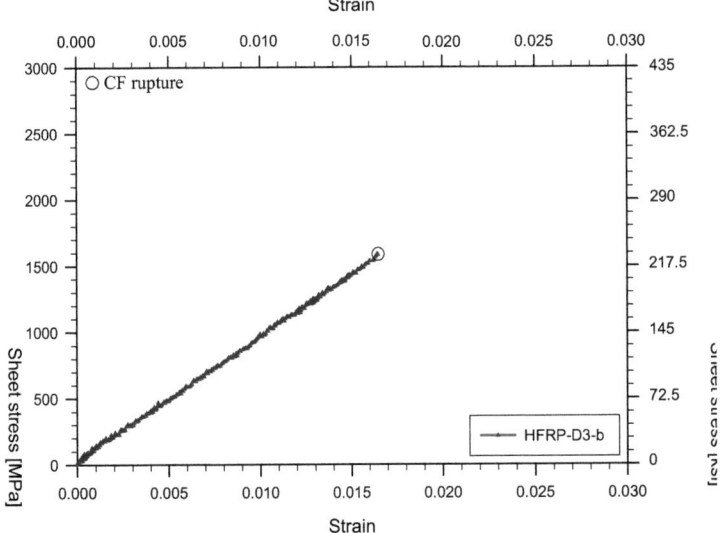

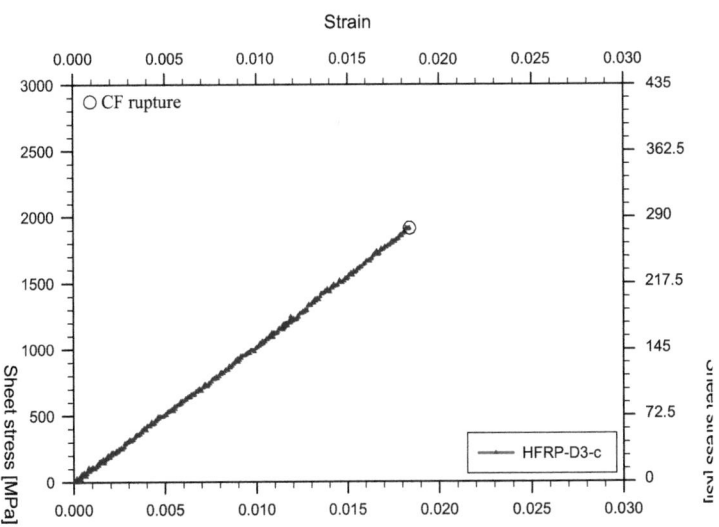

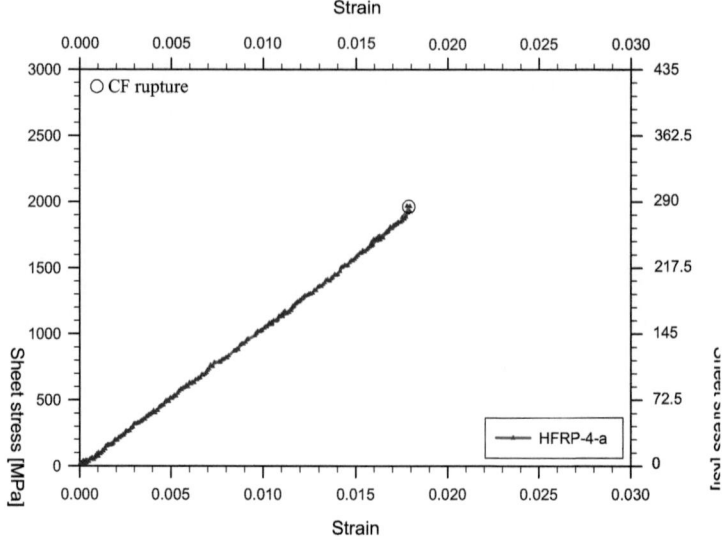

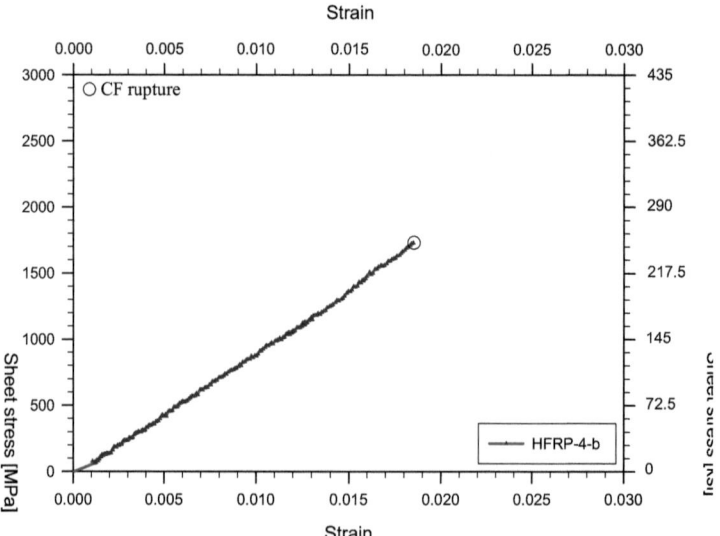

48

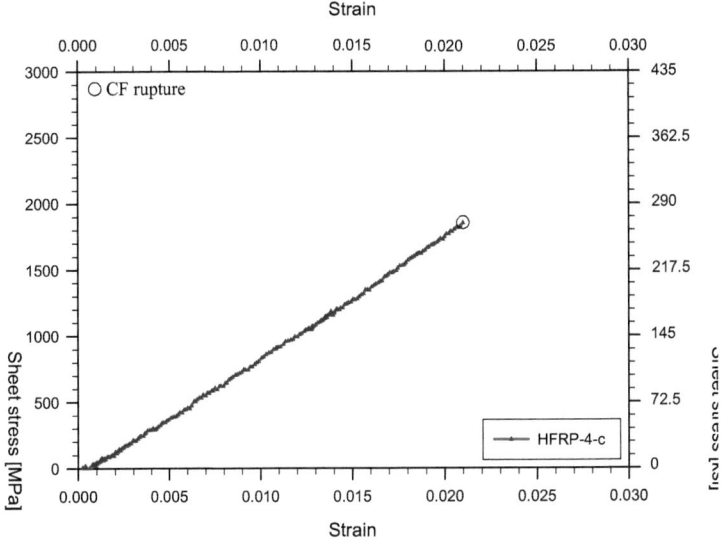

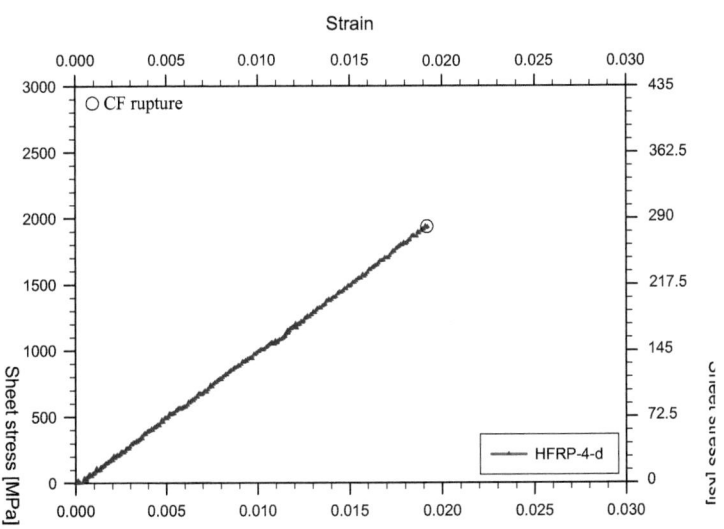

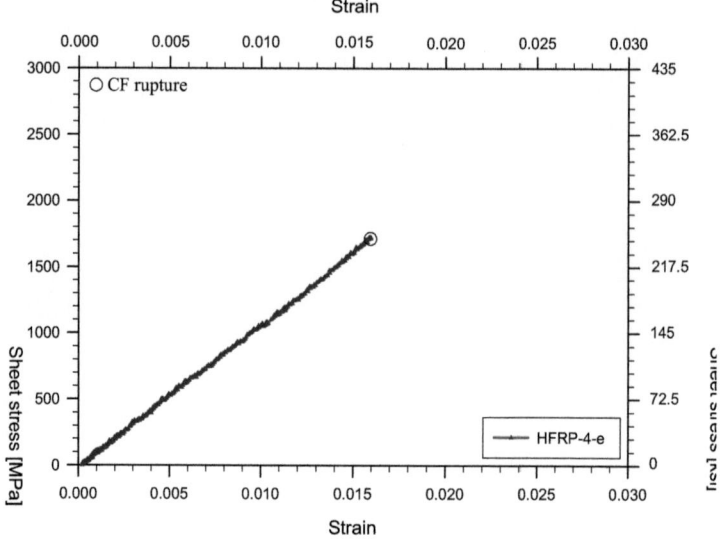

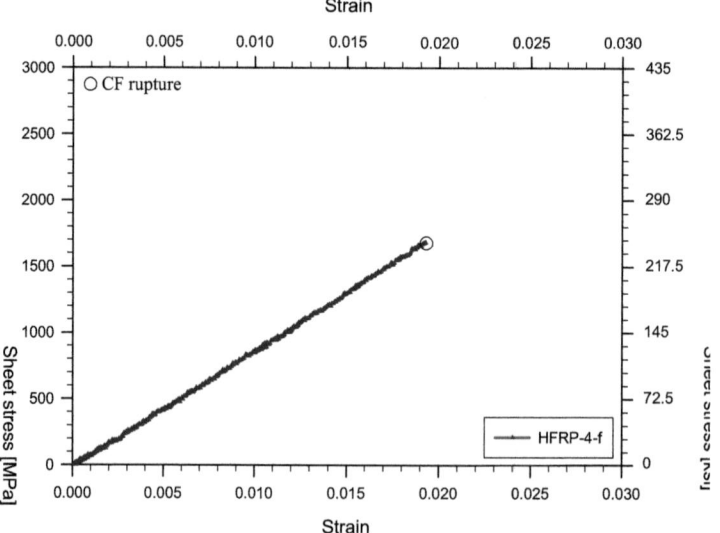

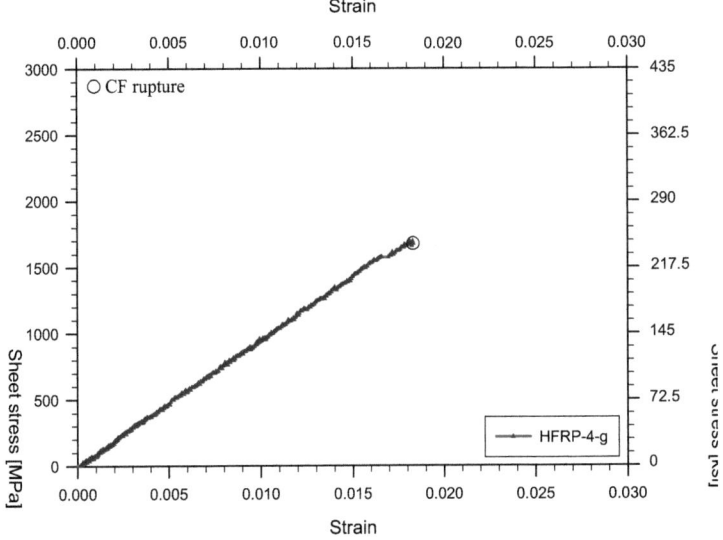

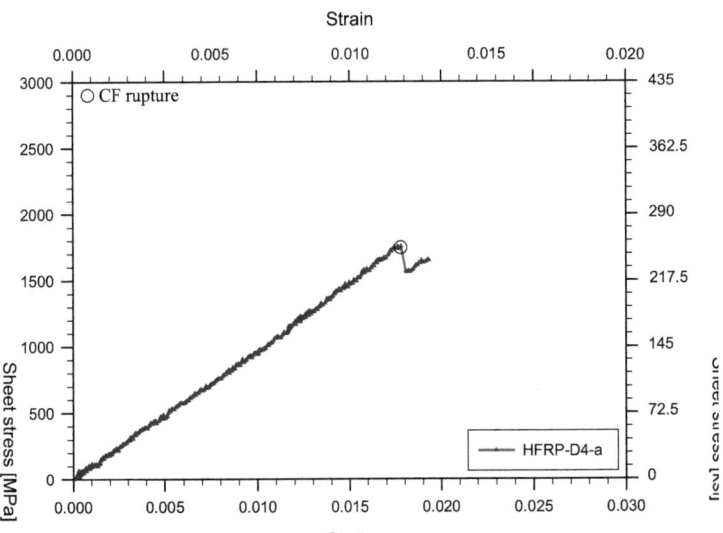

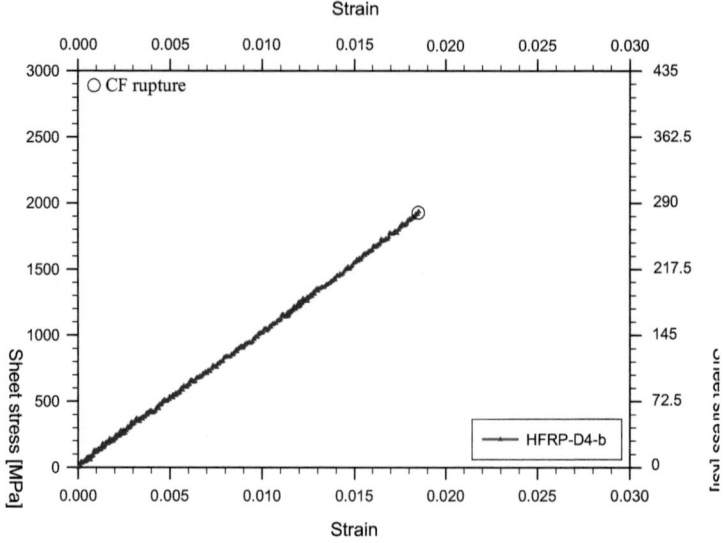

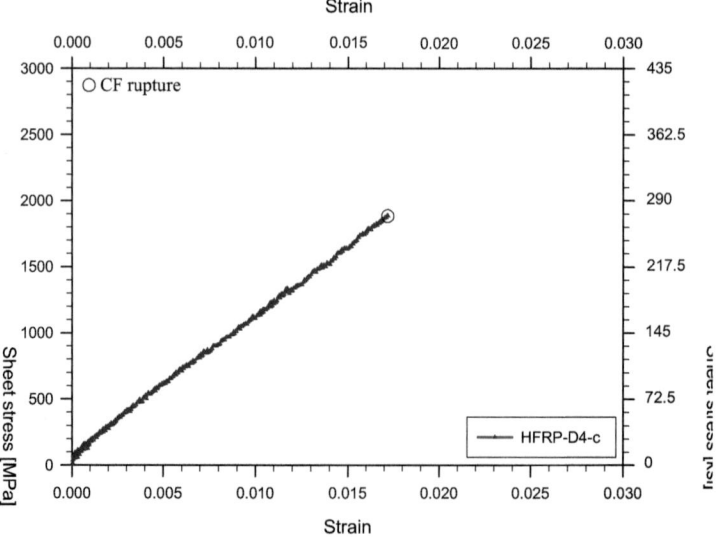

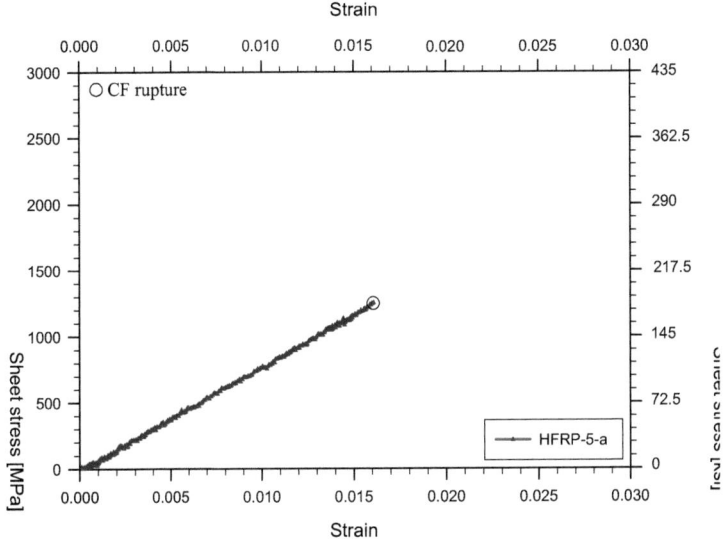

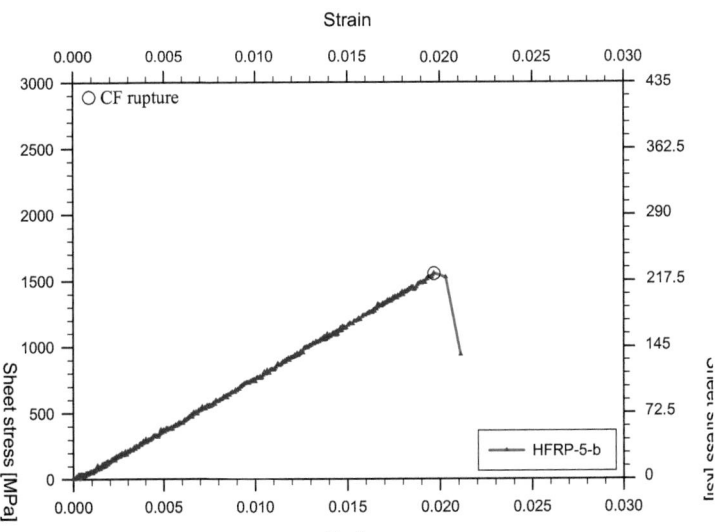

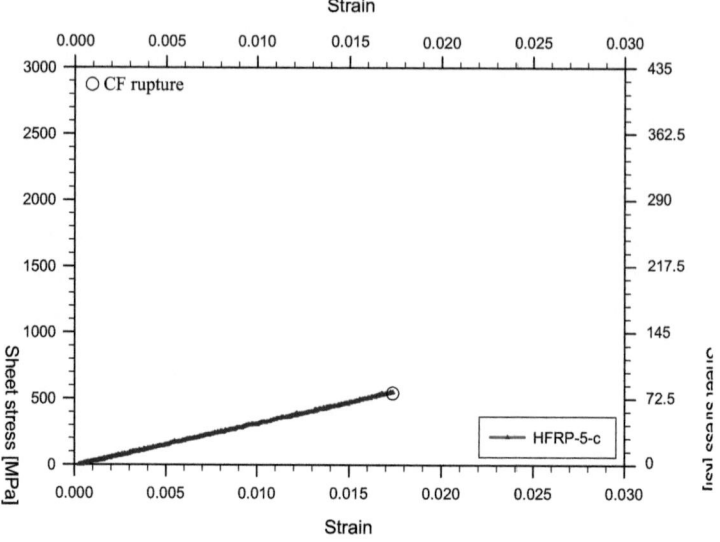

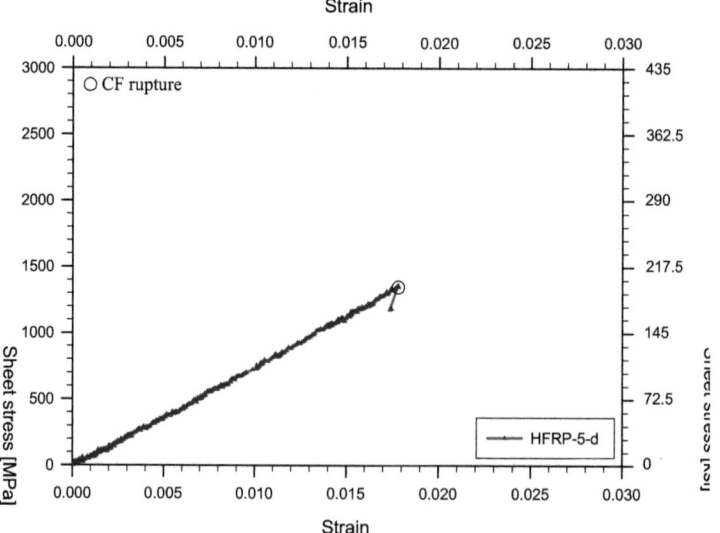

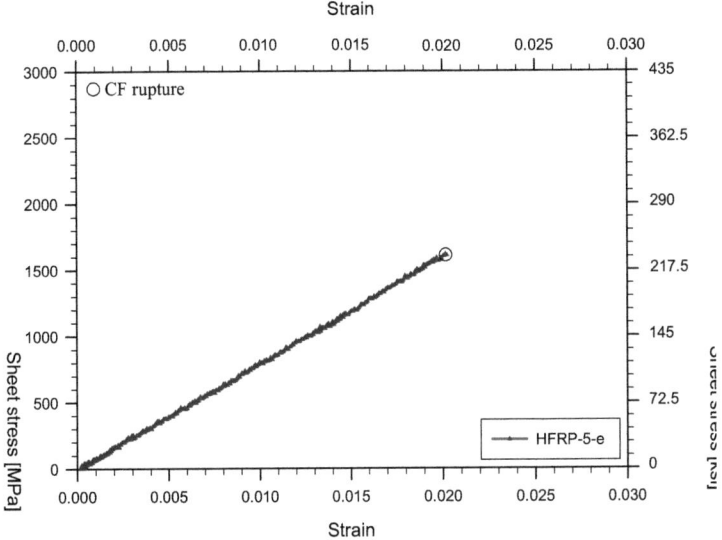

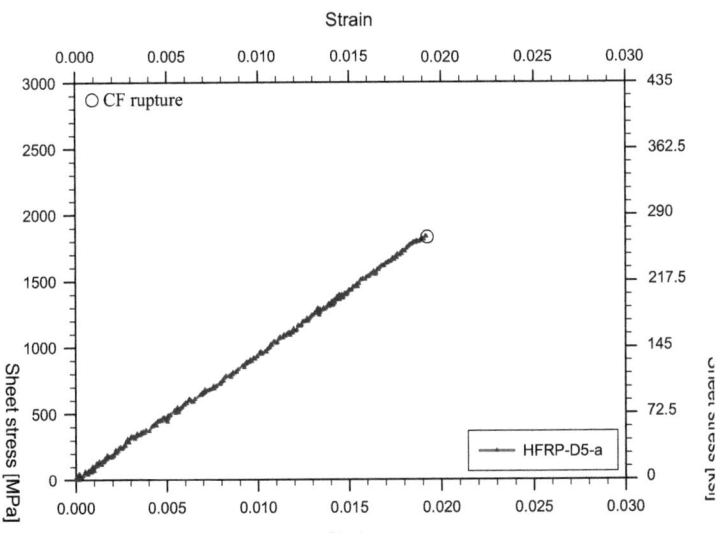

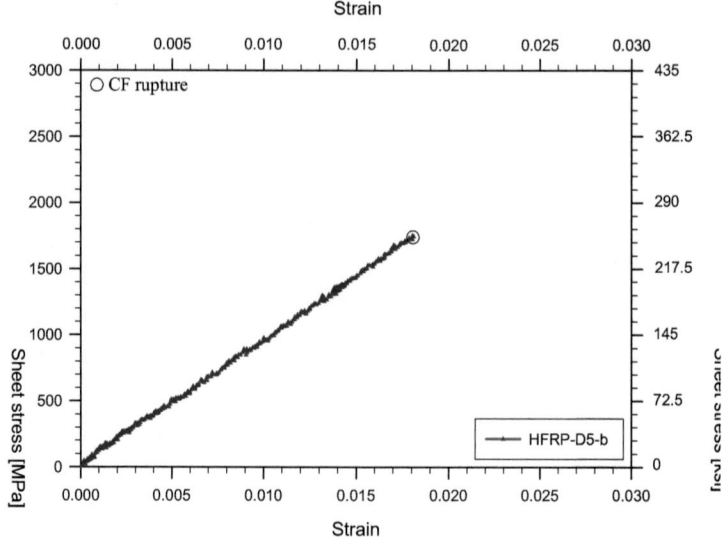

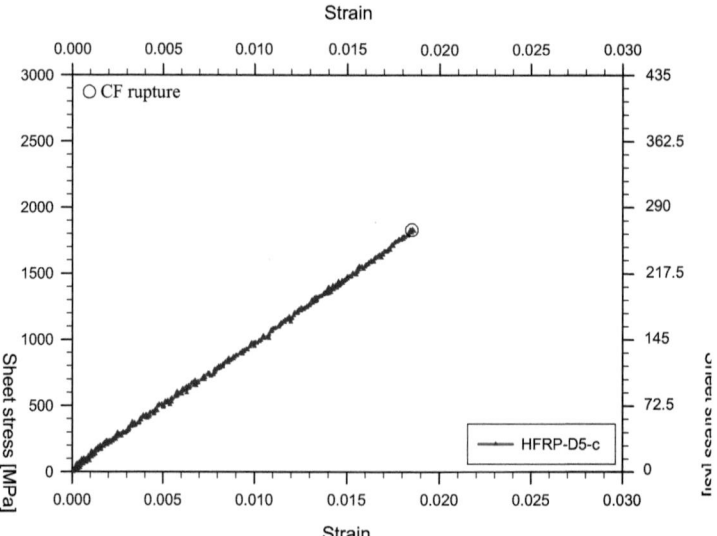

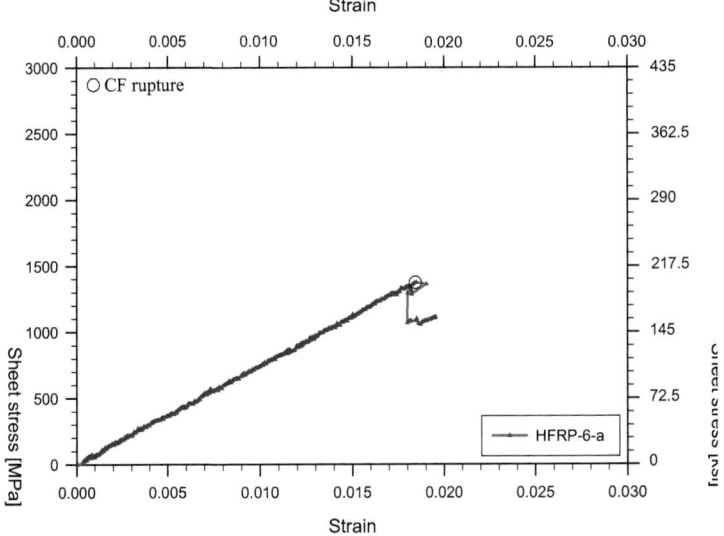

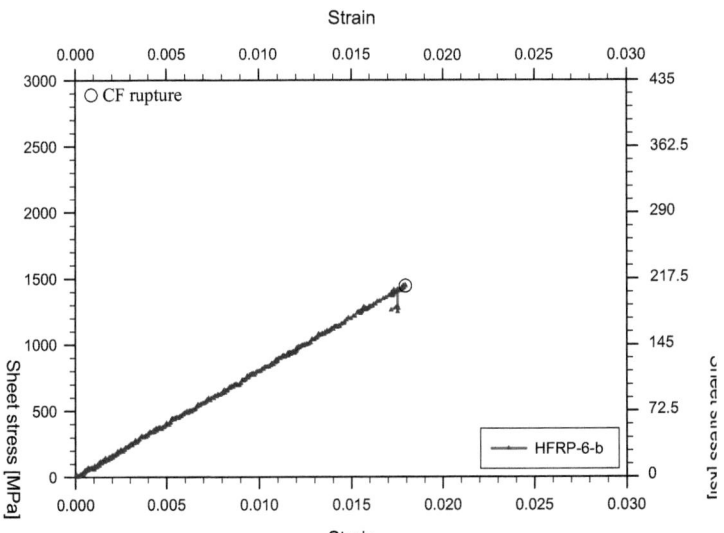

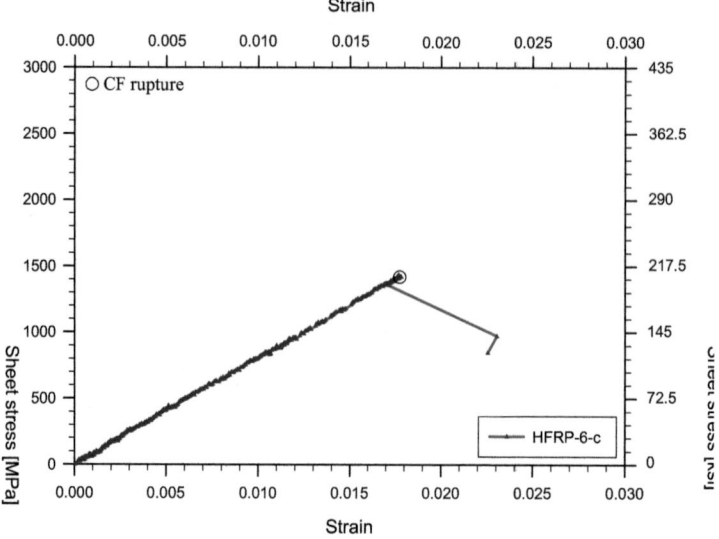

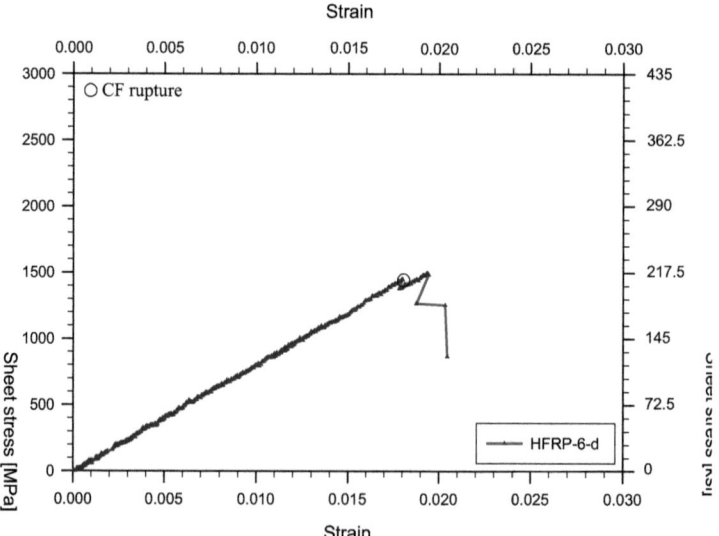

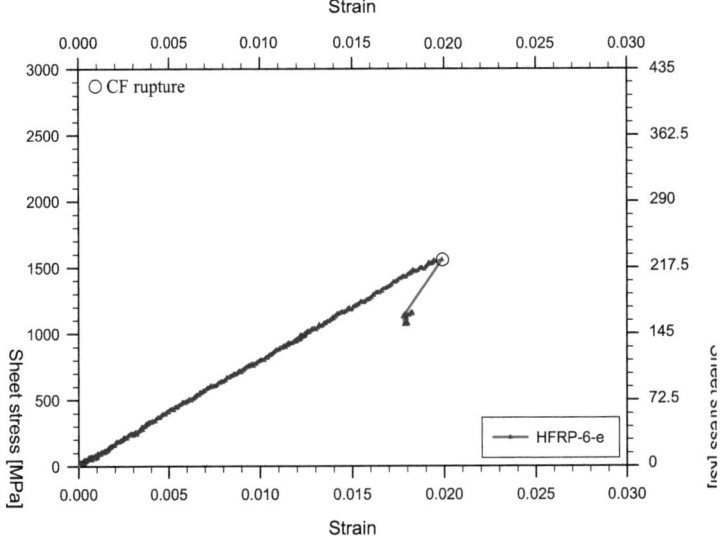

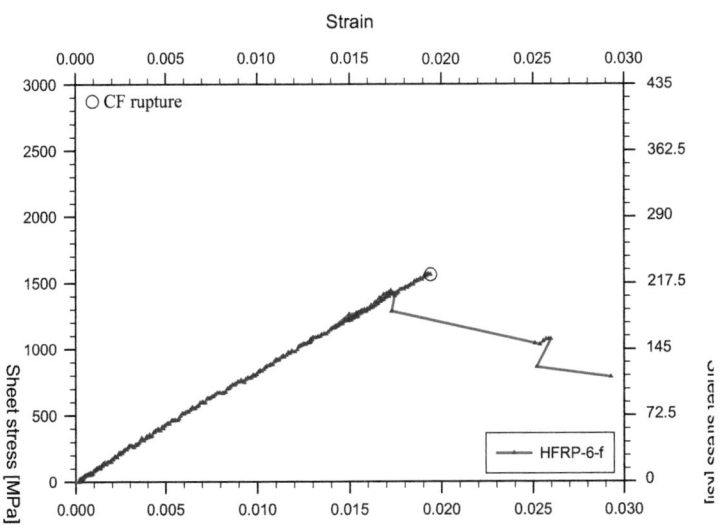

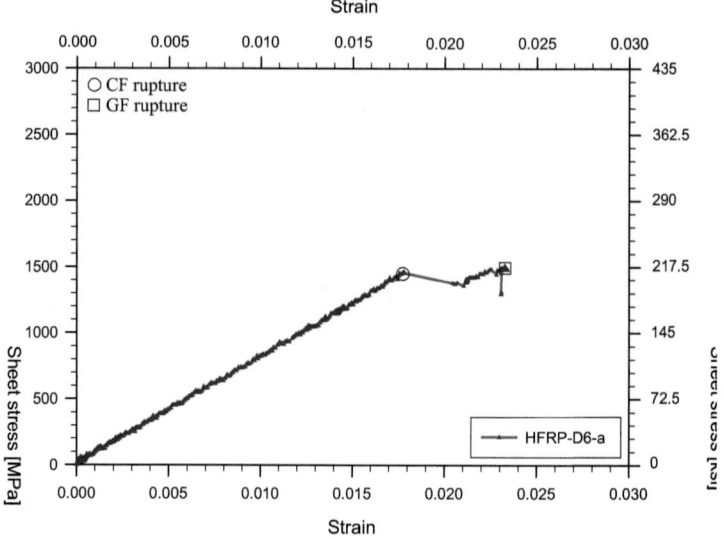

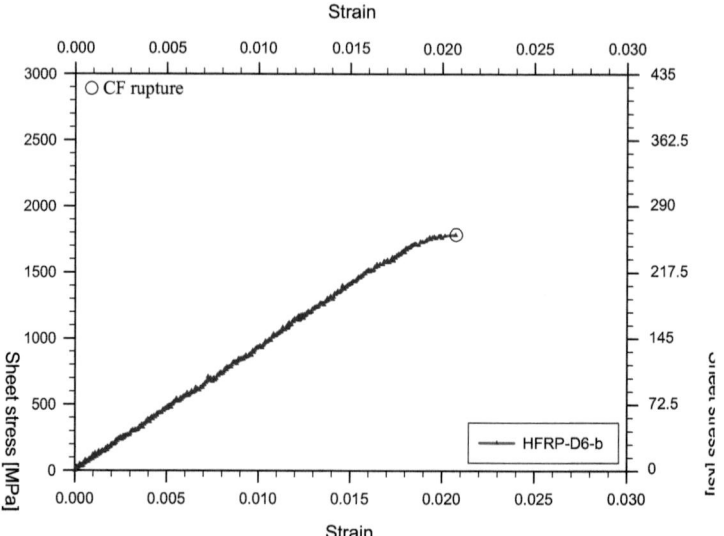

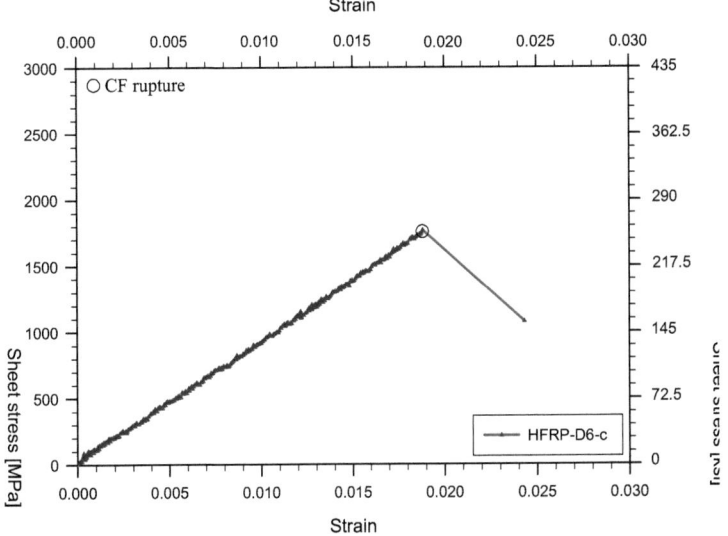

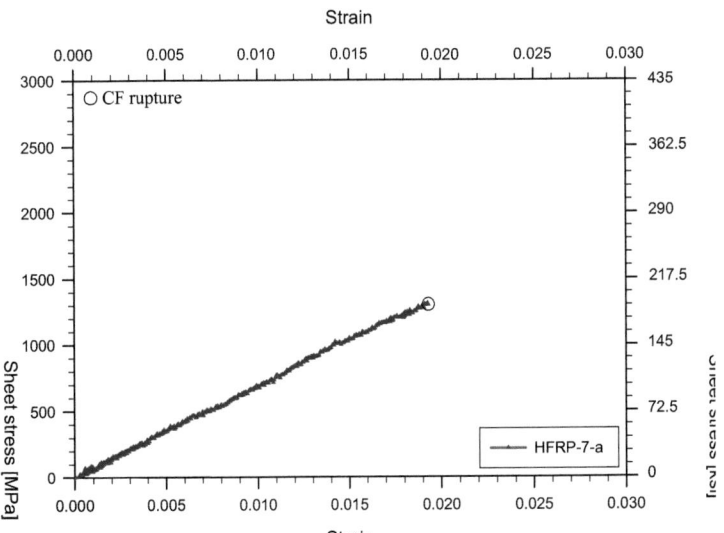

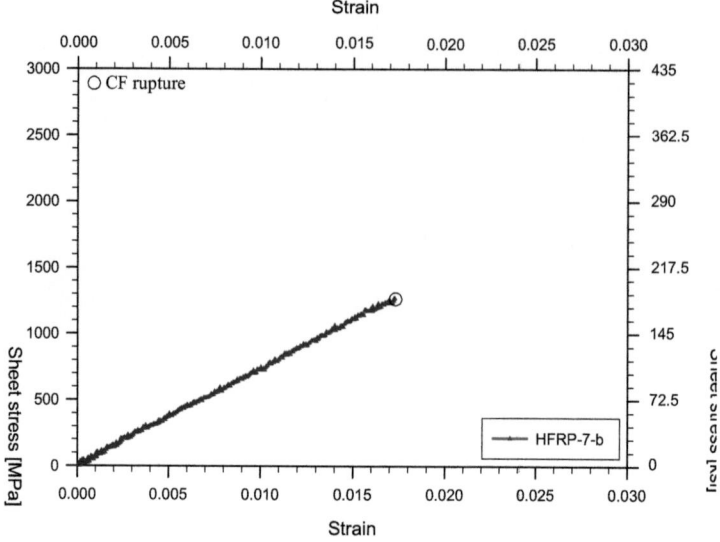

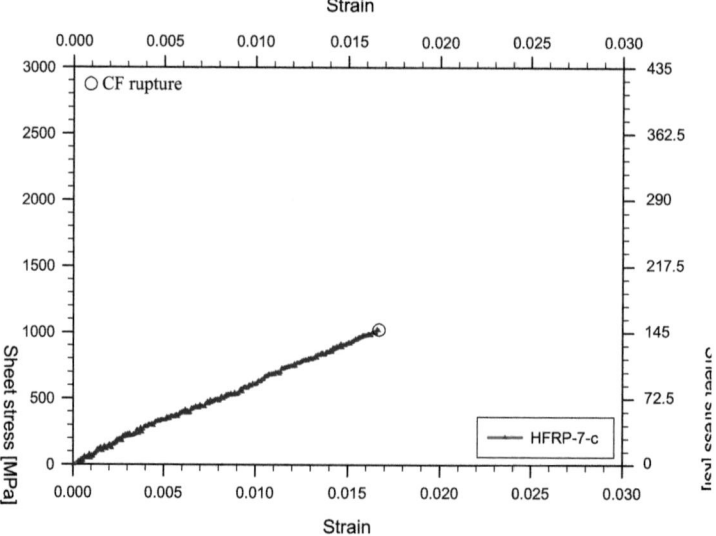

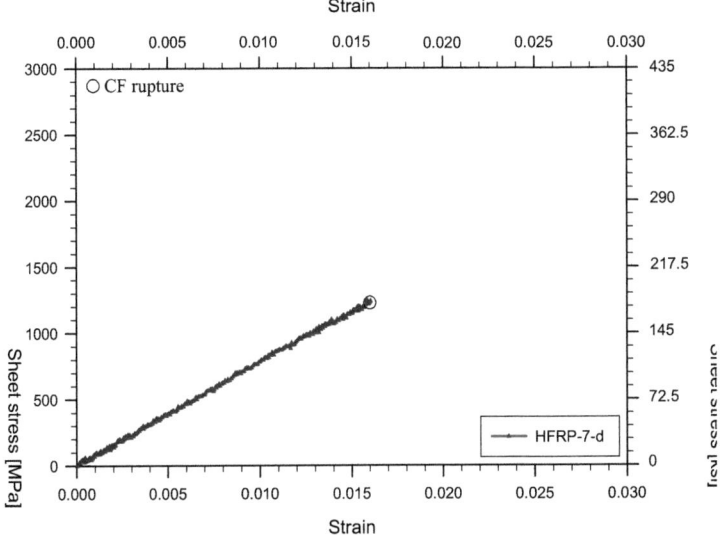

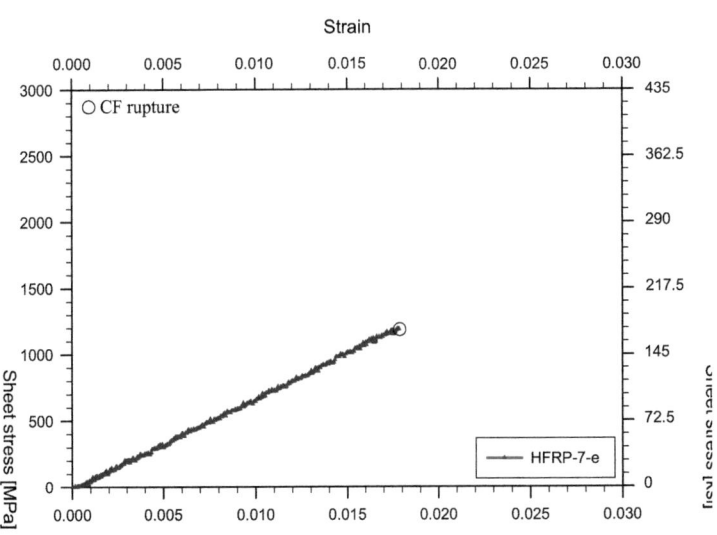

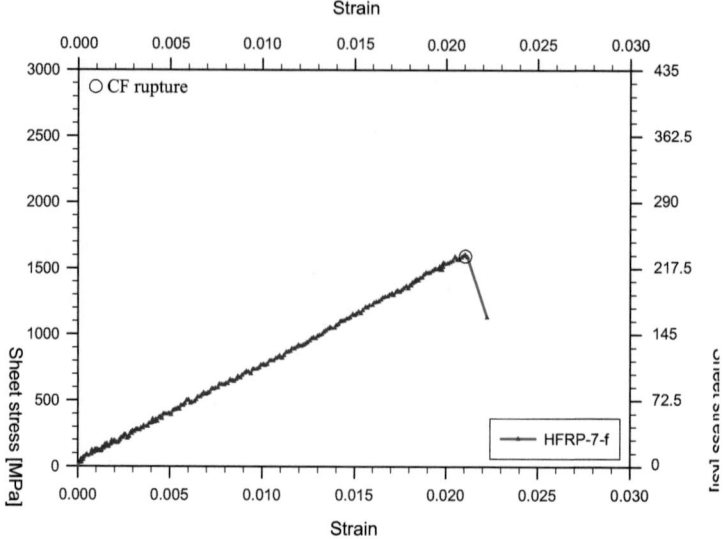

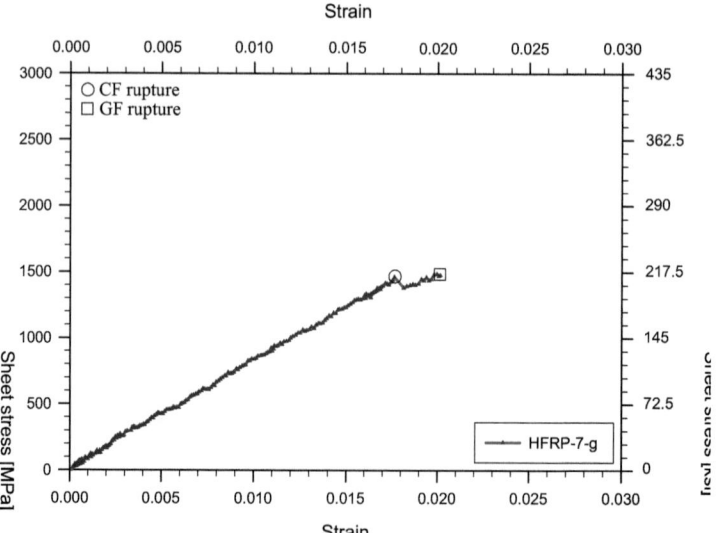

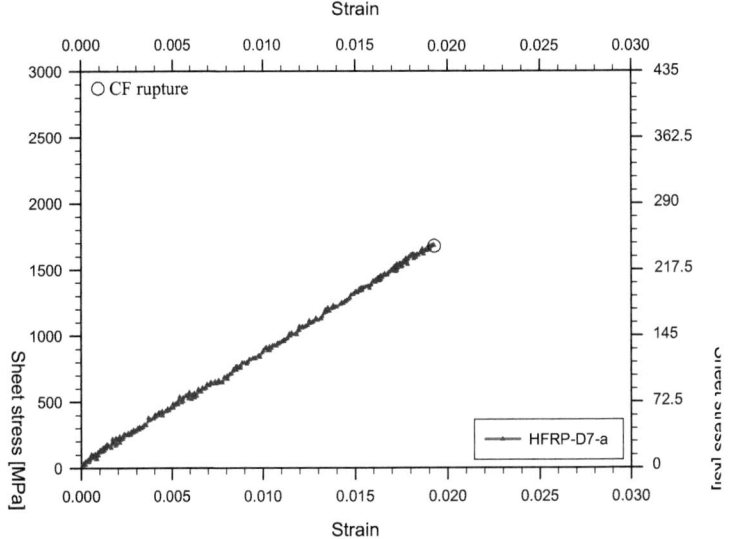

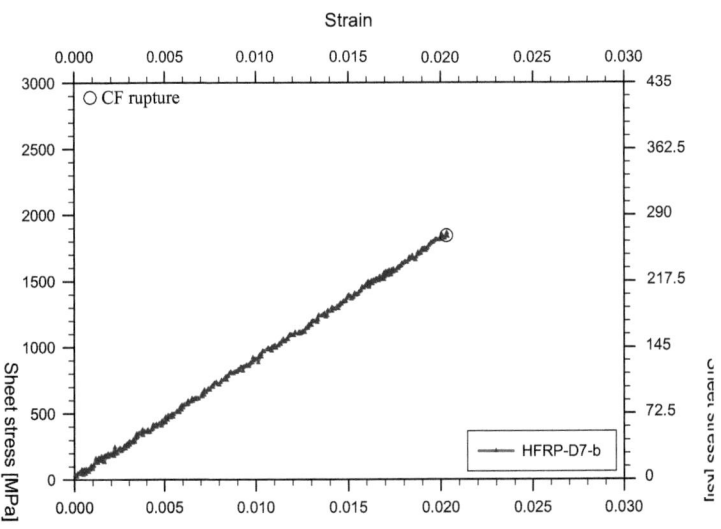

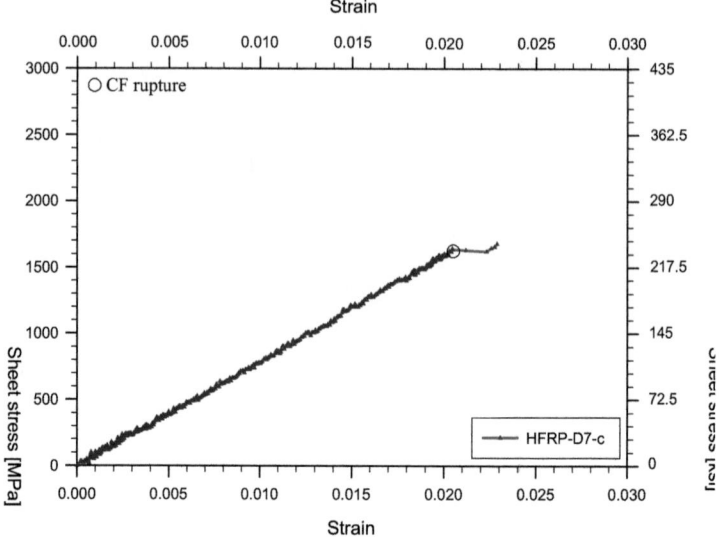

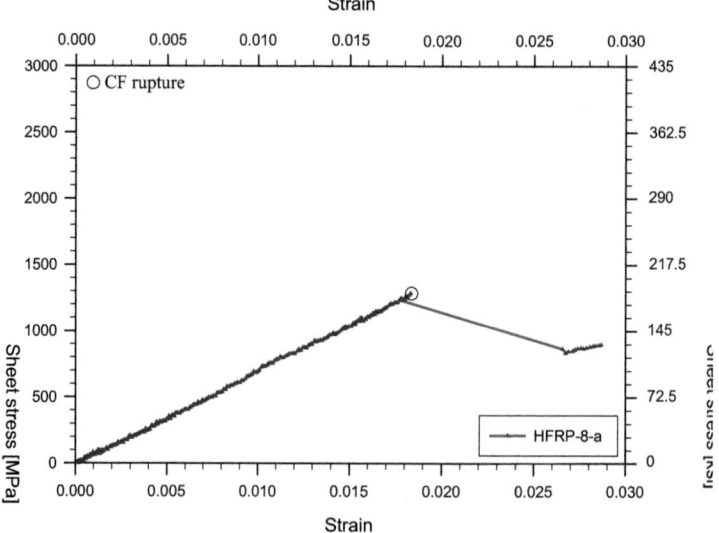

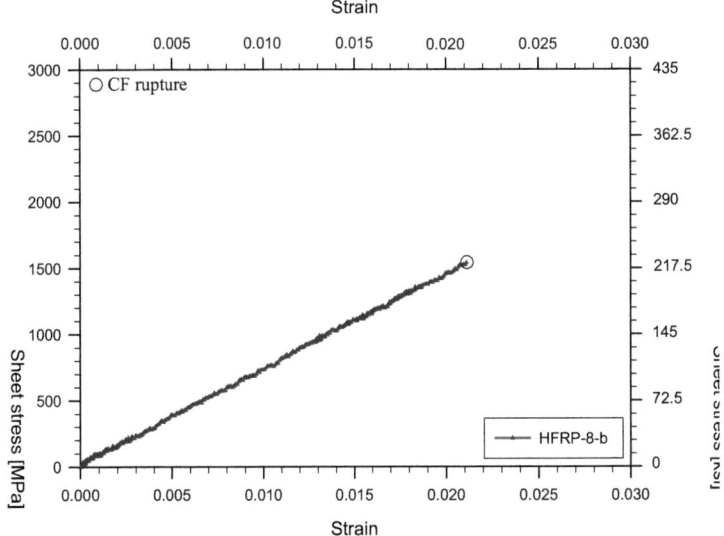

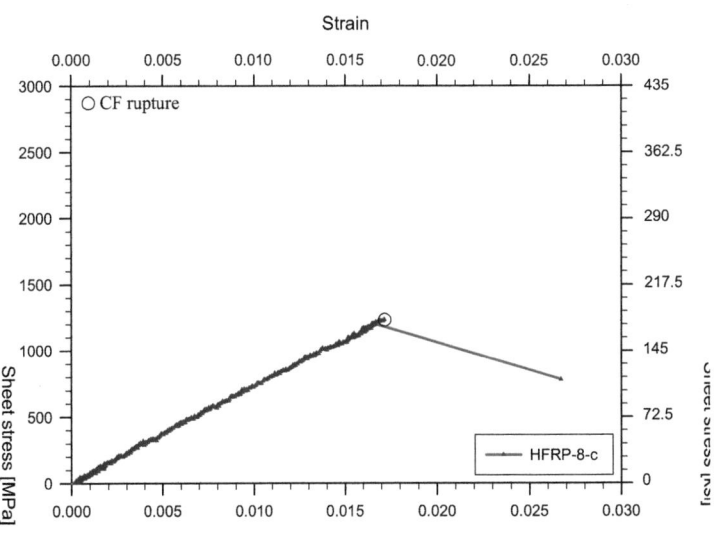

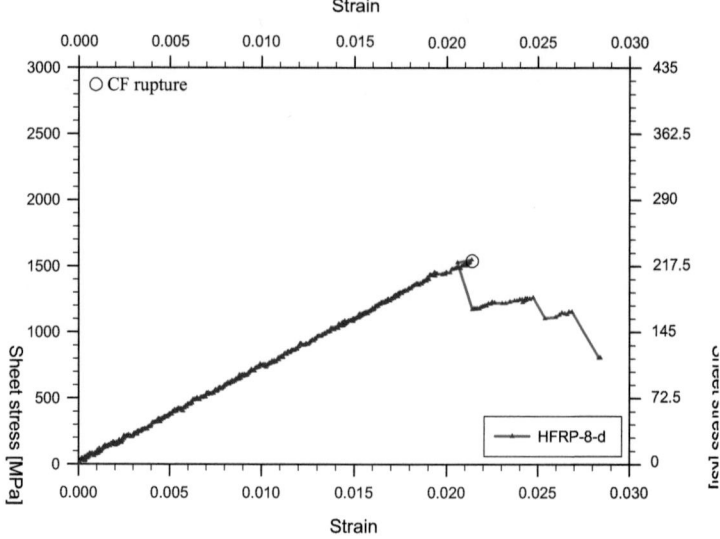

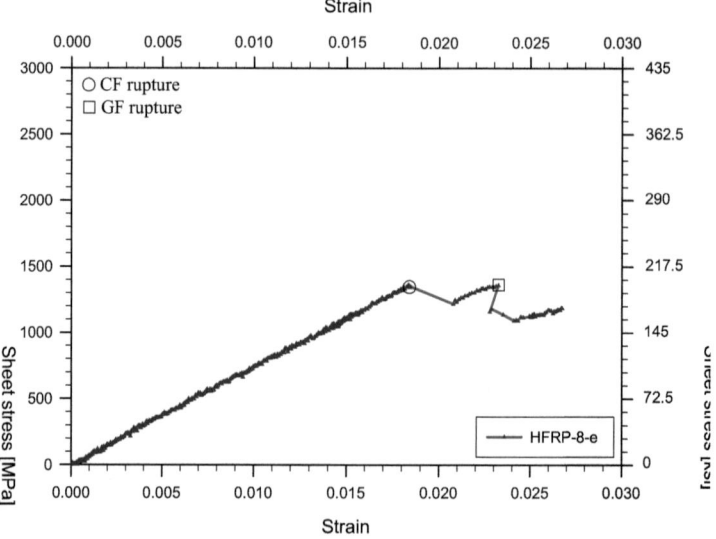

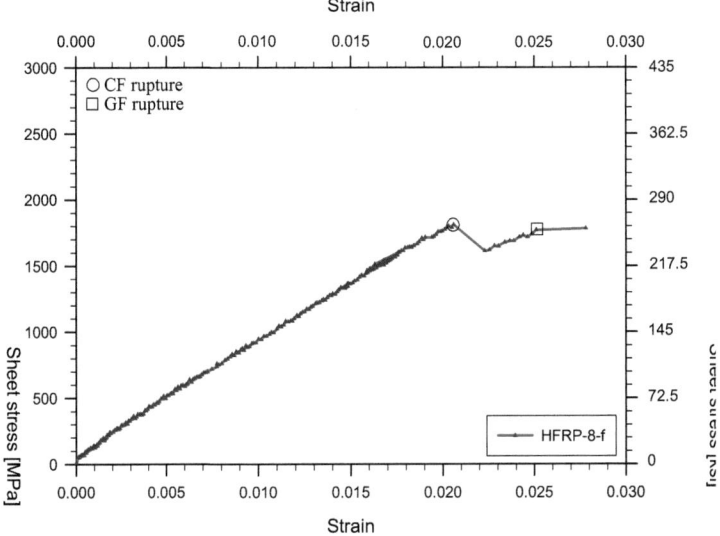

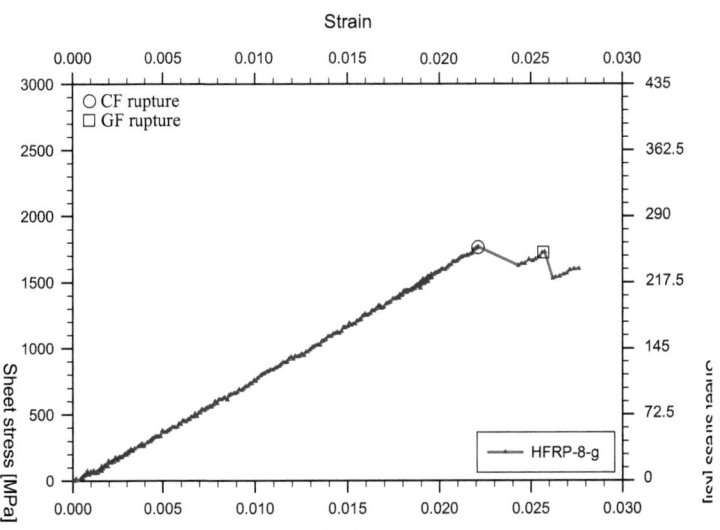

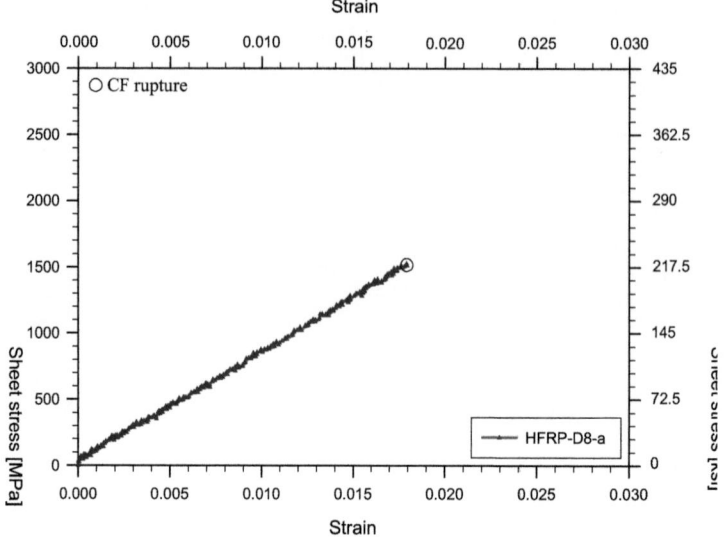

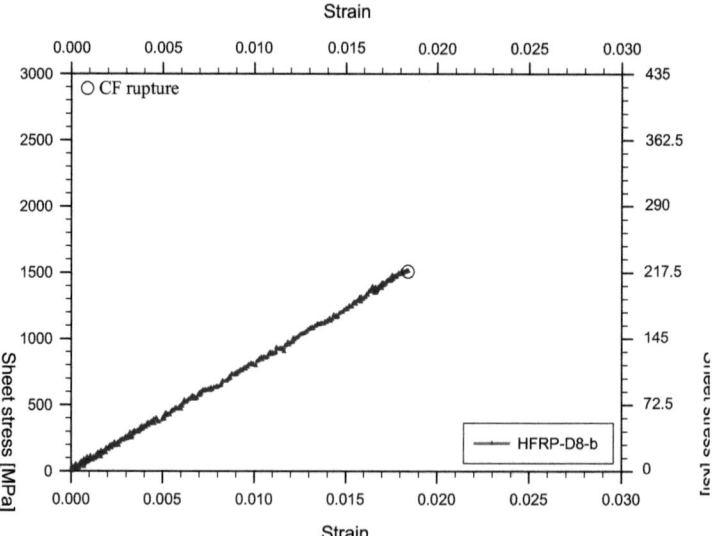

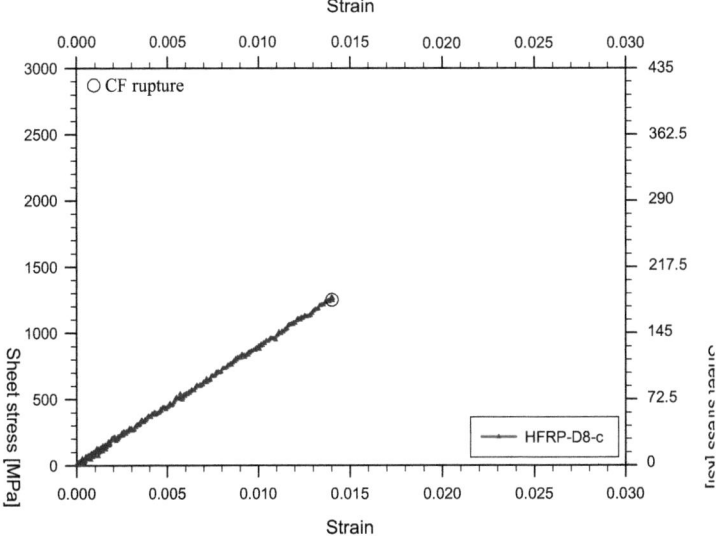

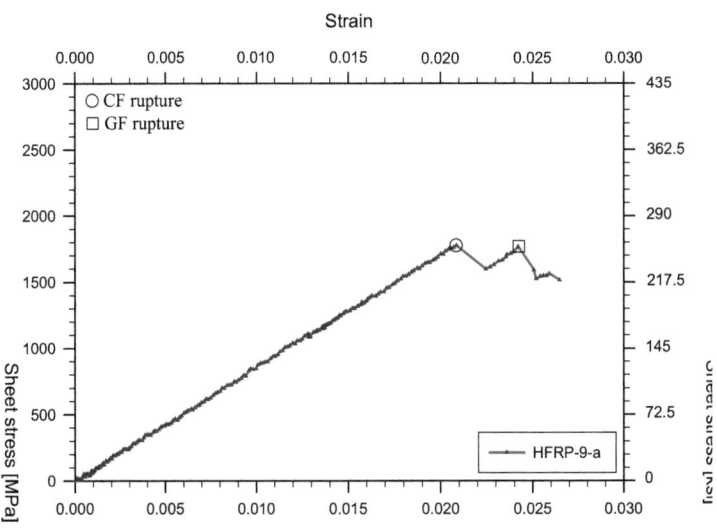

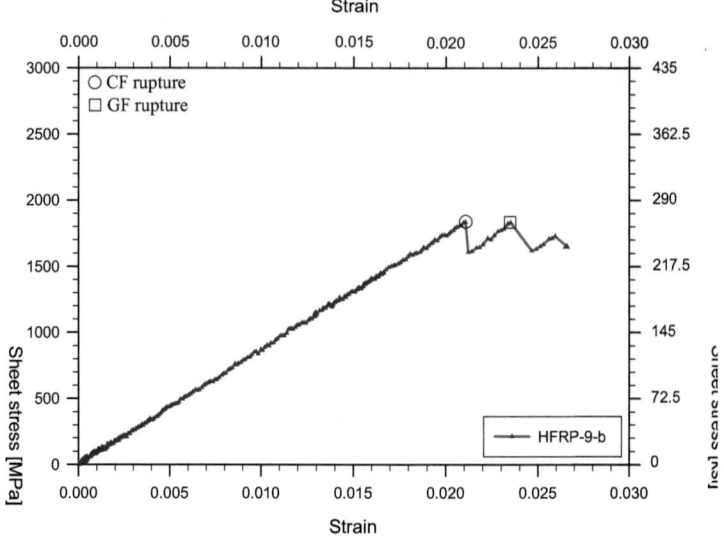

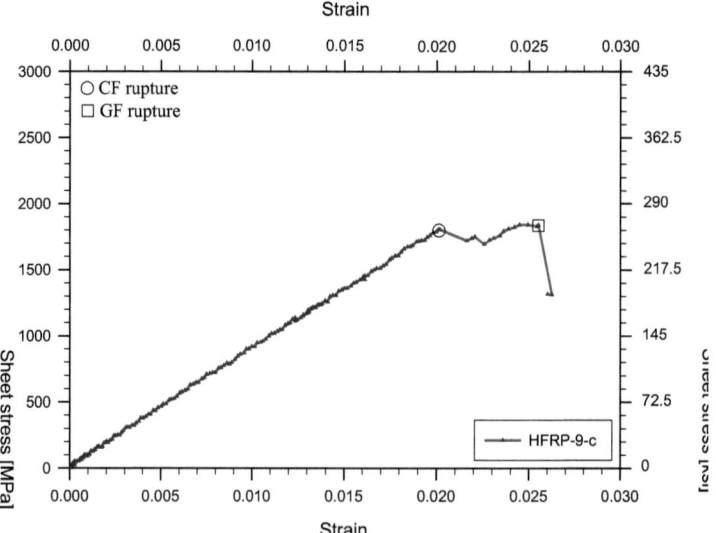

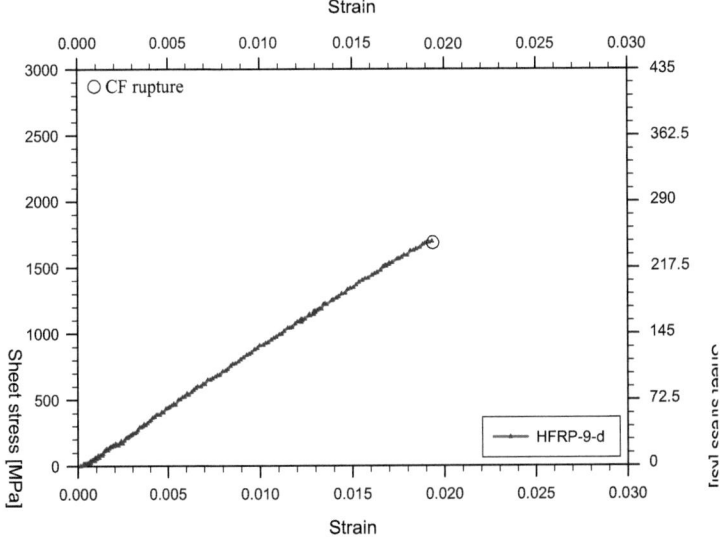

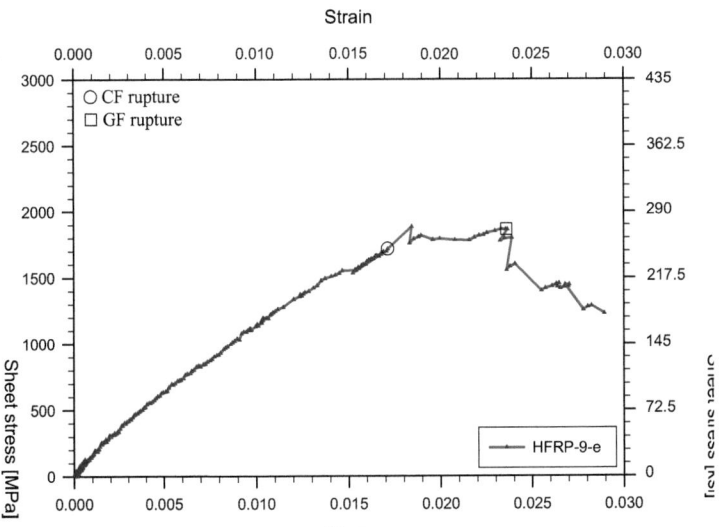

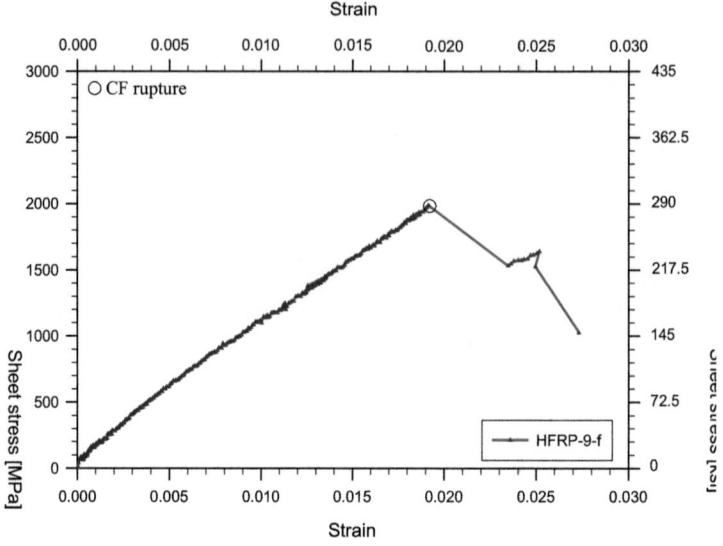

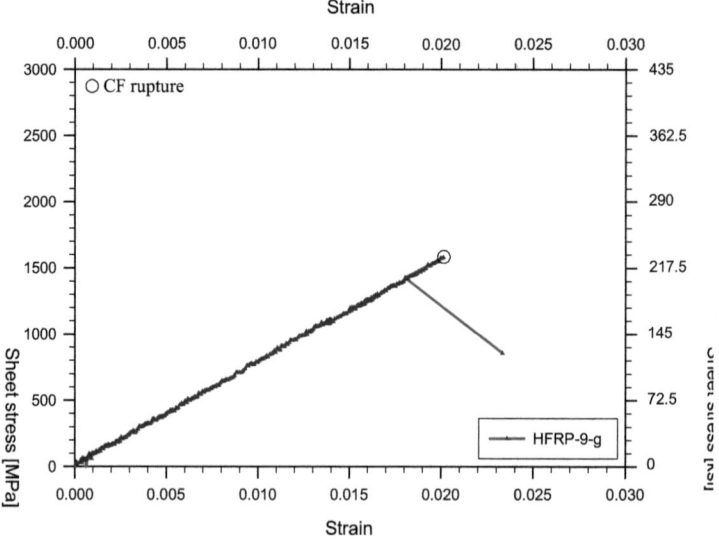

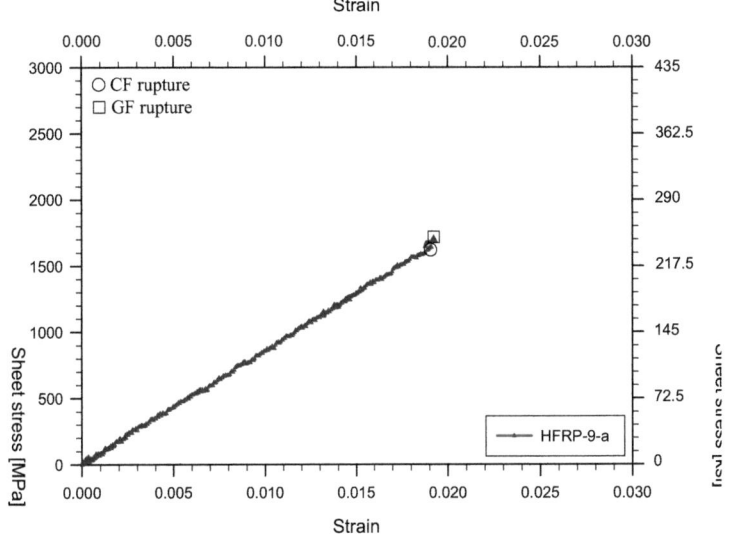

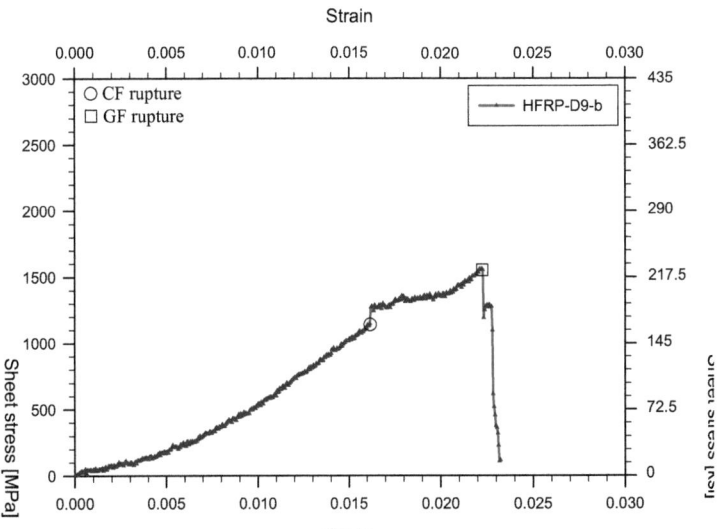

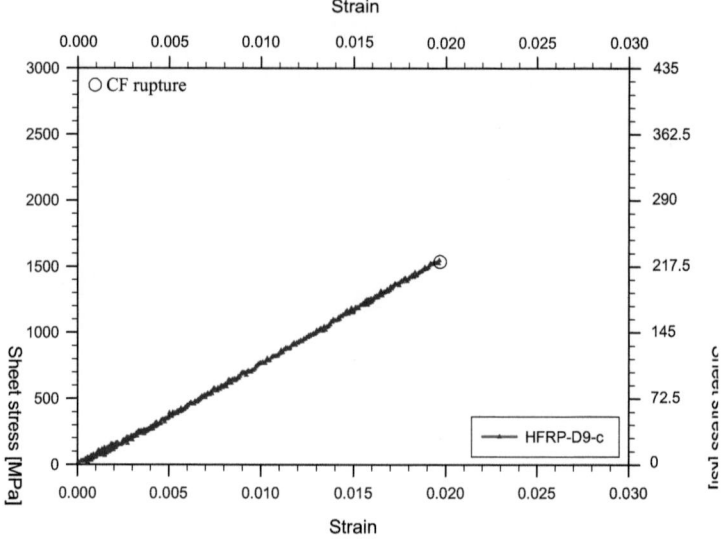

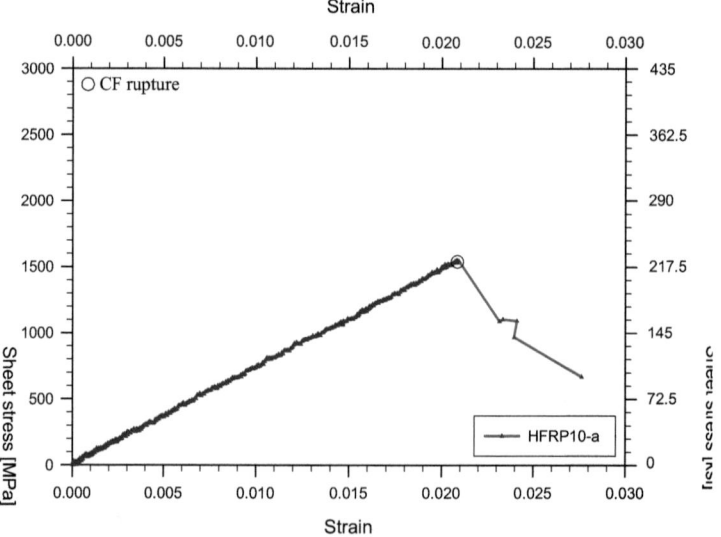

76

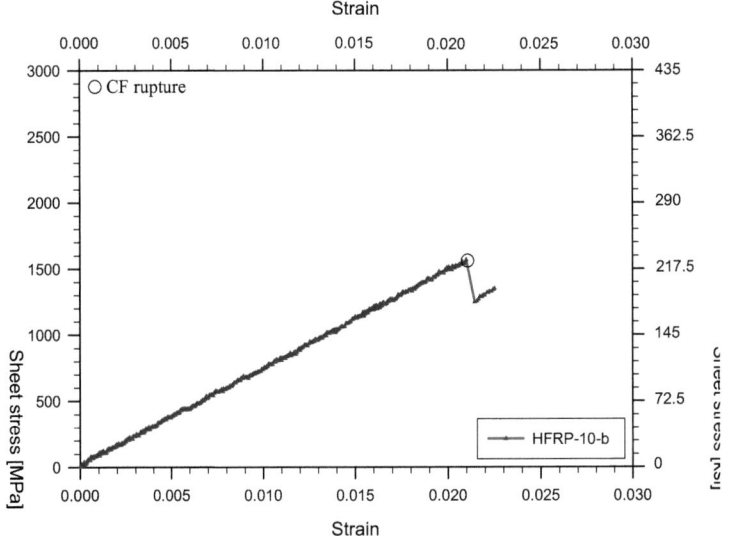

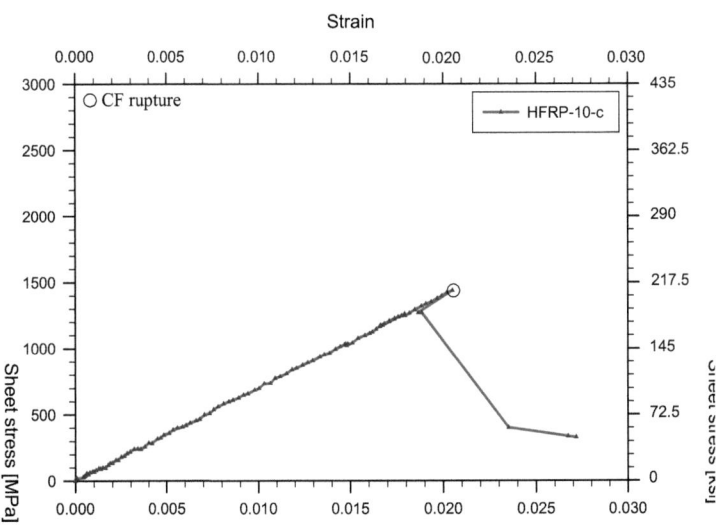

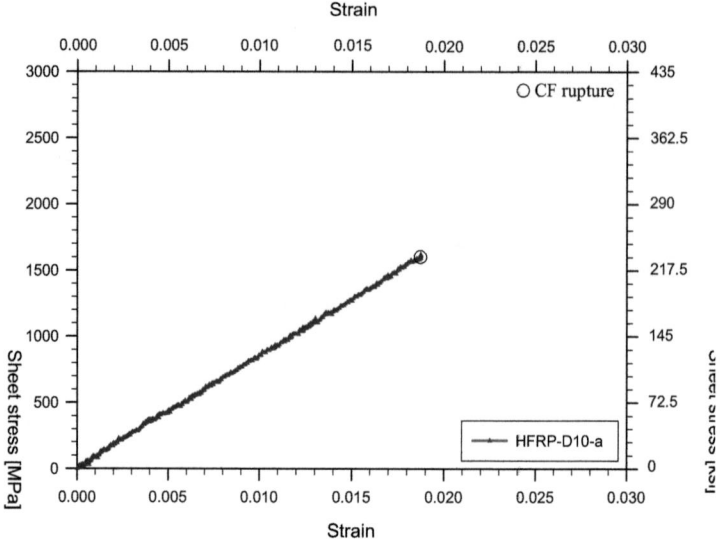

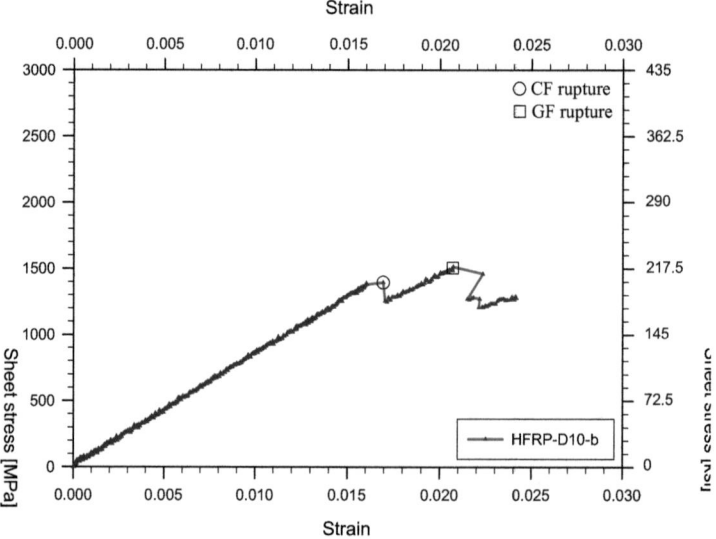

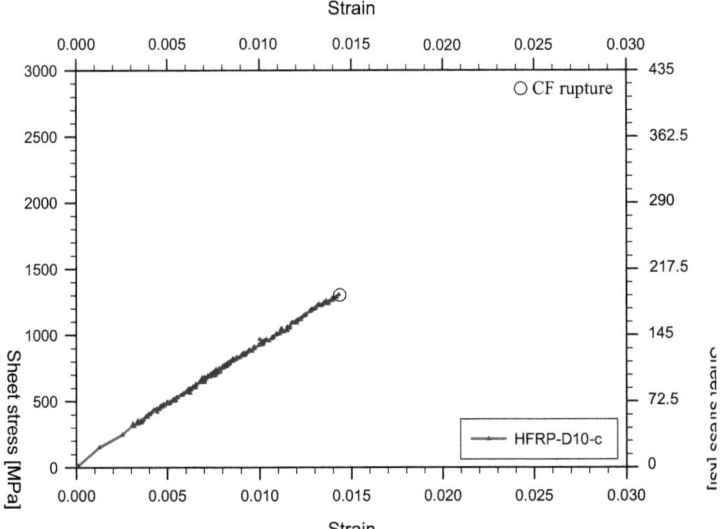

More Books!

yes

I want morebooks!

Buy your books fast and straightforward online - at one of the world's
fastest growing online book stores! Environmentally sound due to
Print-on-Demand technologies.

Buy your books online at

www.get-morebooks.com

Kaufen Sie Ihre Bücher schnell und unkompliziert online – auf einer der am
schnellsten wachsenden Buchhandelsplattformen weltweit!
Dank Print-On-Demand umwelt- und ressourcenschonend produziert.

Bücher schneller online kaufen

www.morebooks.de

OmniScriptum Marketing DEU GmbH
Heinrich-Böcking-Str. 6-8
D - 66121 Saarbrücken
Telefax: +49 681 93 81 567-9

info@omniscriptum.com
www.omniscriptum.com

OMNIScriptum

MIX
Papier aus verantwortungsvollen Quellen
Paper from responsible sources
FSC® C105338

Printed by Books on Demand GmbH, Norderstedt / Germany